The Carbon Chain in Carbon Dioxide Industrial Utilization Technologies

A shift towards implementation of renewable energy has disadvantages, such as power availability, storage capacity, and accompanying costs, and therefore the potential of clean fossil fuel technologies to ensure the stability of electricity generation needs to be reconsidered until these challenges will be overcome. These clean technologies can help prevent the greenhouse effect and, at the same time, guarantee energy security, as coal is a widespread, price-stable raw material that is available in large quantities. This book focuses on the carbon chain, starting from the formation of CO_2, through its capture, possible cleaning, to the production of useful products such as dimethylether, methanol, and carbonated cement prefabricates. The comprehensive case study presents the research results of an international team established within the "CCS-CCU technology for carbon footprint reduction using bio-adsorbents" (BIOCO2) project.

The Carbon Chain in Carbon Dioxide Industrial Utilization Technologies

A Case Study

Edited by
Dariusz Wawrzyńczak, Izabela Majchrzak-Kucęba,
Covadonga Pevida, Giuseppe Bonura,
Rita Nogueira and Marcello De Falco

CRC Press
Taylor & Francis Group
Boca Raton London New York

CRC Press is an imprint of the
Taylor & Francis Group, an **informa** business

Designed cover image: © Shutterstock

First edition published 2022
by CRC Press
6000 Broken Sound Parkway NW, Suite 300, Boca Raton, FL 33487-2742

and by CRC Press
4 Park Square, Milton Park, Abingdon, Oxon, OX14 4RN

CRC Press is an imprint of Taylor & Francis Group, LLC

Library of Congress Cataloging-in-Publication Data
Names: Wawrzyńczak, Dariusz, editor.
Title: The carbon chain in carbon dioxide industrial utilization technologies: a case study / edited by Dariusz Wawrzyńczak, Izabela Majchrzak-Kucęba, Covadonga Pevida, Giuseppe Bonura, Rita Nogueira, Marcello De Falco.
Description: First edition. | Boca Raton: CRC Press, 2022. |
Includes bibliographical references and index.
Identifiers: LCCN 2022034120 (print) | LCCN 2022034121 (ebook) |
ISBN 9781032373546 (hbk) | ISBN 9781032373553 (pbk) | ISBN 9781003336587 (ebk)
Subjects: LCSH: Carbon dioxide—Industrial applications—Case studies. |
Catalysis—Case studies. | Carbon dioxide—Recycling—Case studies. |
Carbon sequestration—Case studies. | Carbon dioxide mitigation—Case studies.
Classification: LCC TP244.C1 C345 2023 (print) | LCC TP244.C1 (ebook) |
DDC 363.738/746—dc23/eng/20220901
LC record available at https://lccn.loc.gov/2022034120
LC ebook record available at https://lccn.loc.gov/2022034121

ISBN: 9781032373546 (hbk)
ISBN: 9781032373553 (pbk)
ISBN: 9781003336587 (ebk)

DOI: 10.1201/9781003336587

Typeset in Times New Roman
by codeMantra

Contents

Acknowledgments

This book has been supported by the Polish National Agency for Academic Exchange under Grant No. PPI/APM/2019/1/00042/U/00001

NAWA POLISH NATIONAL AGENCY FOR ACADEMIC EXCHANGE

Editors

Dariusz Wawrzyńczak is an Assistant Professor in the Department of Advanced Energy Technologies, Faculty of Infrastructure and Environment at the Czestochowa University of Technology in Poland. His research interest focuses on carbon dioxide capture and utilization, adsorption technology, solid adsorbent analysis, integration of conventional and renewable technologies, and hydrogen storage. He is an author and co-author of more than 15 publications in Journal Citation Reports (JCR) and three patents, and has given a number of presentations at international conferences. He was a contractor in many national and international projects, which included CO_2 capture research at the laboratory, bench, and pilot scales. He supervised investigations during pilot plant measurement campaigns in real industrial conditions at power plants where the dual-reflux vacuum pressure swing adsorption method was tested.

Professor Izabela Majchrzak-Kucęba is a Dean of the Faculty of Infrastructure and Environment at the Czestochowa University of Technology in Poland. She is a member of the Power Engineering Problems Committee of the Polish Academy of Sciences. Her achievements as author, co-author, and co-editor comprise about 180 scientific and research publications, including two author's monographs, six co-edited monographs, one academic textbook, and articles in journals. Her research interests focus on CCUS technology, pressure swing adsorption technology for CO_2 capture, energy utilization, advanced sorbents (MOFs, micro and mesoporous materials), and thermal analysis. She has been conducting the above-mentioned research activities by carrying out 25 scientific and research projects. She is the co-author of invention patents about the method of carbon dioxide capture. She was the organizer of two CCS Schools for students and doctoral students. She was Chairwoman of the Organizational Committee of the 1st International Conference on Advanced CO_2 Capture Technologies for Clean Coal Energy Generation at Krakow.

Dr. Pevida holds a Research Scientist position at INCAR-CSIC. She obtained an MEng and a PhD in Chemical Engineering from the University of Oviedo, and she was awarded the Young PhD Researcher Award from the Spanish Carbon Group in 2005. She was a research fellow at IRCELyon (CNRS-Université de Lyon 1) and the University of Nottingham and a visiting researcher at the RCCS of Heriot-Watt University (Edinburgh). She currently leads the Energy Processes and Emission Reduction (PrEM) Group at INCAR-CSIC. Her research interests focus on the abatement of CO_2 emissions and on the utilization of biomass to produce renewable H_2. She has co-authored more than 120 publications in Journal Citation Reports (JCR), led national and European R&D projects, and supervised nine PhDs to completion in the topic of CO_2 capture. She was the Head of the Department of Coal, Energy and Environment at INCAR-CSIC between March 2013 and May 2019. She is the leader of the Capture Group of the Spanish CO_2 Technology platform (PTECO2).

Dr. Pevida's expertise regarding CO_2 capture is on the development of solid sorbents, particularly carbon-based, and on adsorption processes for CO_2 separation from gaseous streams.

Giuseppe Bonura is a CNR Senior Researcher at the Institute for Advanced Energy Technologies "Nicola Giordano" in Messina, Italy. He graduated in Industrial Chemistry in 2002, then receiving his PhD in Chemical Processes and Innovative Technologies at the University of Messina in 2009. His scientific interests deal with the development of catalytic systems for the production of clean alternative fuels. He is responsible for several International and National research projects, especially in the frame of CCU technologies and catalytic applications for the production of alternative carbon-neutral fuels.

He is a scientific expert, evaluator, and monitor, regularly assisting the EU Executive Agencies in the context of research and innovation programmes, dealing with topics on Climate and Environment, Energy, and Transport and Mobility. His scientific production includes numerous publications in international peer-reviewed journals, several communications at national and international conferences, various technical-scientific reports related to contractual activities and the filing of one international patent. He is also an editor of the book *Sustainable Catalysis for Biorefineries* (Green Chemistry Series) published by RSC Books, acting also as an editorial board member and guest editor of various thematic journals in the field of chemical processes and innovative industrial processes.

Rita Nogueira is an Assistant Professor in the Department of Civil Engineering of IST, teaching four courses in Construction Materials and Construction Technology to Integrated Masters of Civil Engineering and Architecture. She is also a member of the CERIS Research and Development Unit, hosted by IST-UL. Her research field is in construction materials, with a focus on mixture design and behavior of mortars and concrete, compatible materials for ancient masonry, conservation of built heritage, and inspection and diagnosis of anomalies. She currently participates in three R&D projects where she develops work in innovative solutions for sustainable construction, namely low CO_2 binders and CO_2 sequestration by cementitious materials and their early-age properties. She has published 15 scientific articles in international peer review journals. Currently, she supervises various final theses of master students and PhD students, who are investigating carbonation and recycling of cement waste.

Marcello De Falco is a full Professor at the Faculty of Engineering at the Campus Bio-Medico of Rome and the University of Rome Tor Vergata. He is a research expert on reactor modeling, energy systems and storage, chemical and solar plant optimization, and selective membranes. He has published more than hundred works in international journals and he was the editor of six books for Springer and Wiley.

He is head of the Process Engineering Research Unit at the Faculty of Engineering at the Campus Bio-Medico of Rome.

Contributors

José Alexandre Bogas
Instituto Superior Técnico
Civil Engineering Research and Innovation for Sustainability
Lisbon, Portugal

Giuseppe Bonura
Consiglio Nazionale delle Ricerche
Istituto di Tecnologie Avanzate per l'Energia "Nicola Giordano"
Messina, Italy

Catia Cannilla
Consiglio Nazionale delle Ricerche
Istituto di Tecnologie Avanzate per l'Energia "Nicola Giordano"
Messina, Italy

Mauro Capocelli
Università Campus Bio-Medico di Roma
Faculty of Engineering
Research Unit "Process Engineering"
Rome, Italy

Marcello De Falco
Università Campus Bio-Medico di Roma
Faculty of Engineering
Research Unit "Process Engineering"
Rome, Italy

Francesco Frusteri
Consiglio Nazionale delle Ricerche
Istituto di Tecnologie Avanzate per l'Energia "Nicola Giordano"
Messina, Italy

María Victoria Gil
Consejo Superior de Investigaciones Científicas
Instituto de Ciencia y Tecnología del Carbono
Oviedo, Spain

Izabela Majchrzak-Kucęba
Czestochowa University of Technology
Faculty of Infrastructure and Environment,
Department of Advanced Energy Technologies
Częstochowa, Poland

Gianluca Natrella
University of Genoa
Department of Naval, Electrical, Electronics and Telecommunications Engineering
Genoa, Italy

Rita Nogueira
Universidade de Lisboa
Instituto Superior Técnico
Civil Engineering Research and Innovation for Sustainability
Lisbon, Portugal

Covadonga Pevida
Consejo Superior de Investigaciones Científicas
Instituto de Ciencia y Tecnología del Carbono
Oviedo, Spain

Paulina Popielak
Czestochowa University of Technology
Faculty of Infrastructure and Environment
Department of Advanced Energy Technologies
Częstochowa, Poland

Nausika Querejeta
Consejo Superior de Investigaciones Científicas
Instituto de Ciencia y Tecnología del Carbono
Oviedo, Spain

María Rodríguez
Consejo Superior de Investigaciones Científicas
Instituto de Ciencia y Tecnología del Carbono
Oviedo, Spain

Fernando Rubiera
Consejo Superior de Investigaciones Científicas
Instituto de Ciencia y Tecnología del Carbono
Oviedo, Spain

André Silva
Universidade de Lisboa
Instituto Superior Técnico
Civil Engineering Research and Innovation for Sustainability
Lisbon, Portugal

Serena Todaro
Università della Calabria
Dipartimento di Ingegneria dell'Ambiente
Territorio e Ingegneria Chimica
Rende, Italy

Dariusz Wawrzyńczak
Czestochowa University of Technology
Faculty of Infrastructure and Environment
Department of Advanced Energy Technologies
Częstochowa, Poland

Preface

The goal of climate policy implemented nowadays is to keep the temperature increase on the Earth below 1.5°C by reduction of greenhouse gas emissions to the atmosphere. This is of particular importance in the case of the energy sector – the most CO_2 emission intensive – based on fossil fuels, especially coal. The reduction is realized by starting the emissions trading system which successively covers more and more industries. The system is imposing additional costs related to carbon dioxide emission, while promoting renewable energy sources this way. Unfortunately, photovoltaic and wind energies show their disadvantage along with an increase in share in the energy mix. Their low availability, due to their dependence on weather conditions, and the inability to efficiently store the energy generated in the period of low demand require the supply of missing amount of energy by other sources. In turn, the fast shift to low-emission fuels may lead to a sharp increase in their prices, and thus increase in electricity production costs. Therefore, the clean fossil fuel technologies for the production of electricity and other CO_2 emission-intensive industries, for example cement, should become more and more popular. These technologies, apart from preventing the greenhouse effect, guarantee energy security at the same time, as coal is a widespread, price-stable raw material that is available in large quantities.

The book comprehensively focuses on selected carbon chain, starting with the formation of CO_2, through its capture, up to the production of useful products such as dimethyl ether, methanol, and carbonated cement prefabricates. The presented case study uses the research results of an international team established within the "CCS-CCU technology for carbon footprint reduction using bio-adsorbents" (BIOCO2) project.

Editors

1 Industrial carbon dioxide capture and utilization technology

A system case study

Dariusz Wawrzyńczak

CONTENTS

1.1 INTRODUCTION

The increase in fossil fuel consumption generates an excessive emission of carbon dioxide, which in 2019 was assessed at 9.9 ± 0.5 GtC/year (9.7 ± 0.5 GtC/year, including the cement carbonation sink) [1]. As a result, in January 2022 the average CO_2 concentration in the atmosphere was equal to 418.19 ppm [2], showing an upward trend, when analyzing the values on a year-to-year basis (Figure 1.1).

The accumulation of CO_2 in the atmosphere leads to an increase in the average temperature of the Earth, which affects climate changes, contributing to heat waves, droughts or floods and extreme winds. Therefore, governments are establishing laws and regulations aimed at the reduction of greenhouse gas emissions. Apart from increasing the efficiency of devices, and switching to low-emission or renewable sources, it is also proposed to capture carbon dioxide and its storage or, what is more acceptable, to apply it, as a useful product in the utilization process. The latter solution is preferred due to the lack of public consent to storing carbon dioxide underground, especially in place of residence.

The most convenient solution is to remove CO_2 emitted from large centralized sources. Based on the available data [4], two sources were selected: the fossil fuel-based power sector and the cement industry (Figure 1.2). These sources provide a

DOI: 10.1201/9781003336587-1

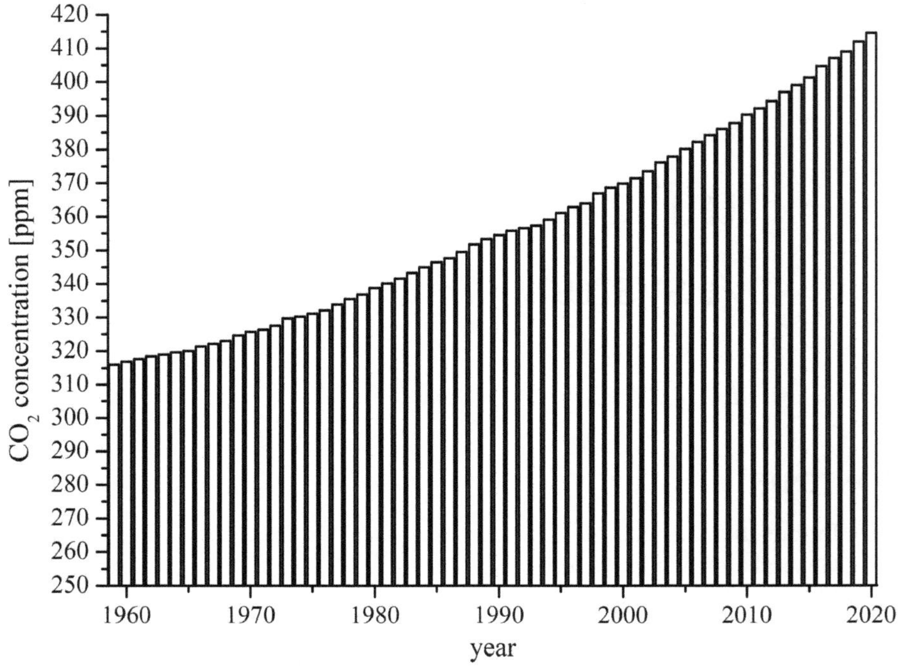

FIGURE 1.1 Changes in the average annual concentration of carbon dioxide in the atmosphere in the years 1959–2020, measured at NOAA Global Monitoring Laboratory [3].

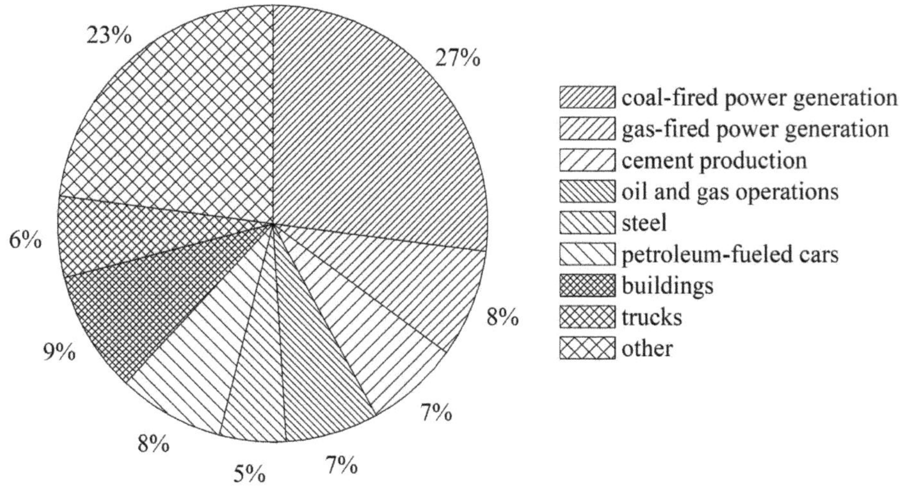

FIGURE 1.2 Greenhouse gas emissions from energy and industrial processes (including methane and nitrous oxide as well as CO_2) [4].

constant flow of diluted CO_2 at a level of about 12 vol% or 22 vol%, which causes the gas stream to have to be enriched to a level required in the utilization process.

The idea presented in this book is to use captured CO_2 as a reactant in fuel production, like dimethyl ether (DME), as well as for the carbonation of concrete products. For this purpose, two process systems in the form of flowcharts have been developed. Chapters 2–6 include the theoretical fundamentals of processes, supplemented by experimental/pilot research data. A case study in the form of process simulation and life cycle assessment is presented in Chapter 7.

1.2 GENERAL SYSTEM APPROACH

Both systems (Figures 1.3 and 1.4) show the entire carbon chain, starting from the capture of CO_2 which is enriched in the adsorption process (Chapter 3) based on the carbon adsorbent coming from biomass processing (Chapter 4) and possible further purification of CO_2, depending on the requirements of the process, as well as its use as a reactant (Chapter 2) for:

a. Fuel production in a single-step hydrogenation process (Chapter 5). Conversion to DME was adopted due to its higher energy density compared to methanol (8.2 kWh/kg versus 6.1 kWh/kg), lower toxicity and potential for being conventionally stored and transported using existing infrastructure and technologies [5]. The properties of DME that make it applicable can be very wide – from aerosol propulsion solvents and coolants, fuel for domestic heating purposes, to fuel for engines, but also as an alternative fuel for direct-feed fuel cells [5, 6].
b. Bonding in materials containing cement (cement prefabricates), as a concrete carbonation method (Chapter 6). Considering the fact that only part of the carbon dioxide emitted during cement production is used for carbonation (the share depends on the cement binder content in the product) [7] and that the process is not conducted continuously (whereby the demand for CO_2 is not constant), the excess of carbon dioxide is used to produce DME fuel (as described in par. a).

Both systems start from a carbon dioxide source, which is a power plant (Figure 1.3) or a cement plant (Figure 1.4). As the carbon dioxide is significantly diluted in the flue gas, CO_2 separation is required to obtain a concentrated reactant for subsequent processes.

Due to the lack of requirements concerning the CO_2 purity level for the DME production process, restrictions have been adopted, which are defined for three applications, namely: pipeline transportation, geological storage and enhanced oil recovery [8], for which the concentration of carbon dioxide in the product must be higher than 95%. In the case of concrete carbonation, it is possible to use flue gas, but this gives CO_2 absorption results worse than those for using pure carbon dioxide – for the concentration of 10–11 vol% the CO_2 the uptake is equal to 6.3% [9]; with 25 vol% CO_2 this value is 9.7% [7], while with pure CO_2 – it ranges from 9.8% [9] to 16% [7], depending on the cement content, carbonation temperature, pressure and exposure time [7]. In addition, due to the CO_2 uptake capacity of the cement-based material

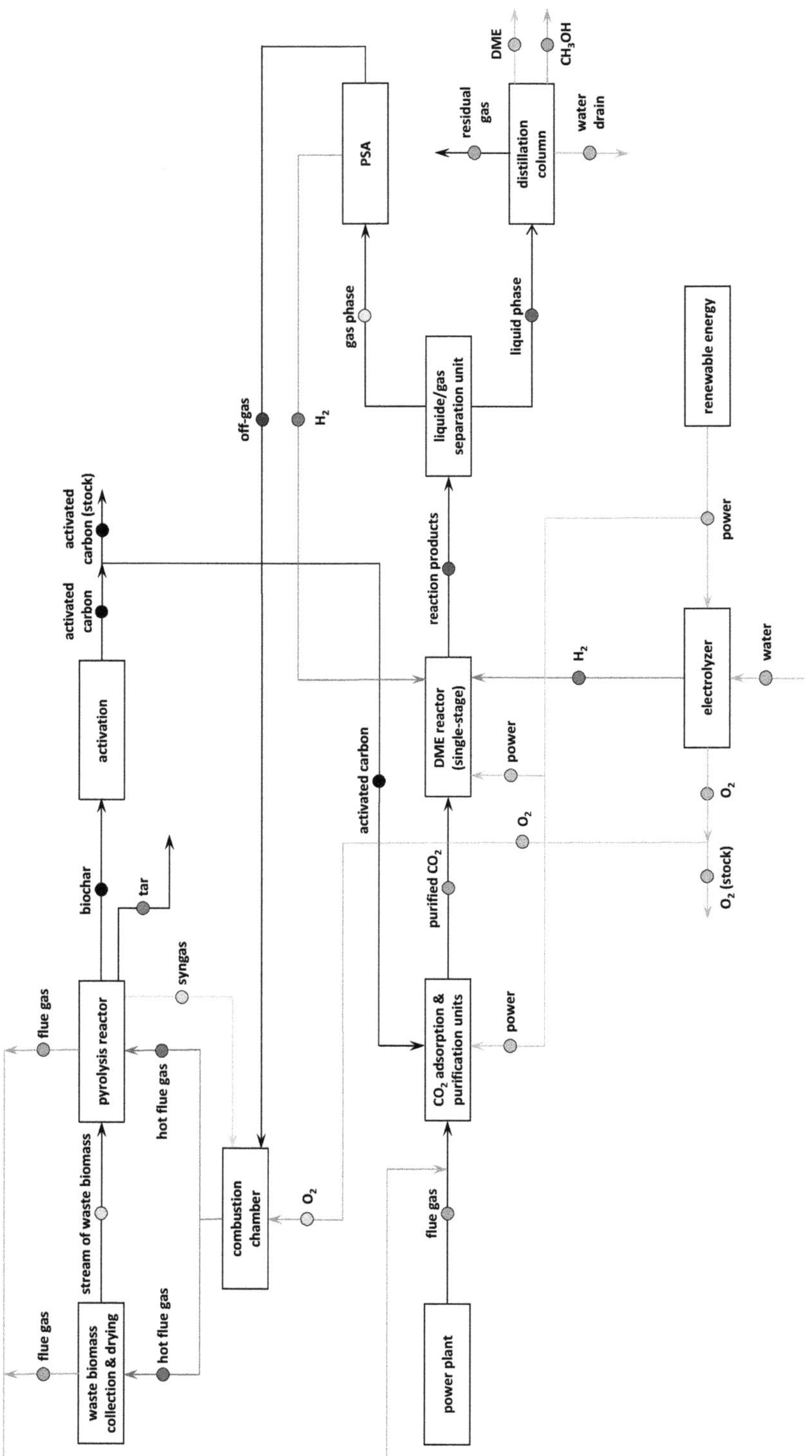

FIGURE 1.3 System of CO_2 conversion into fuel – a power plant case study.

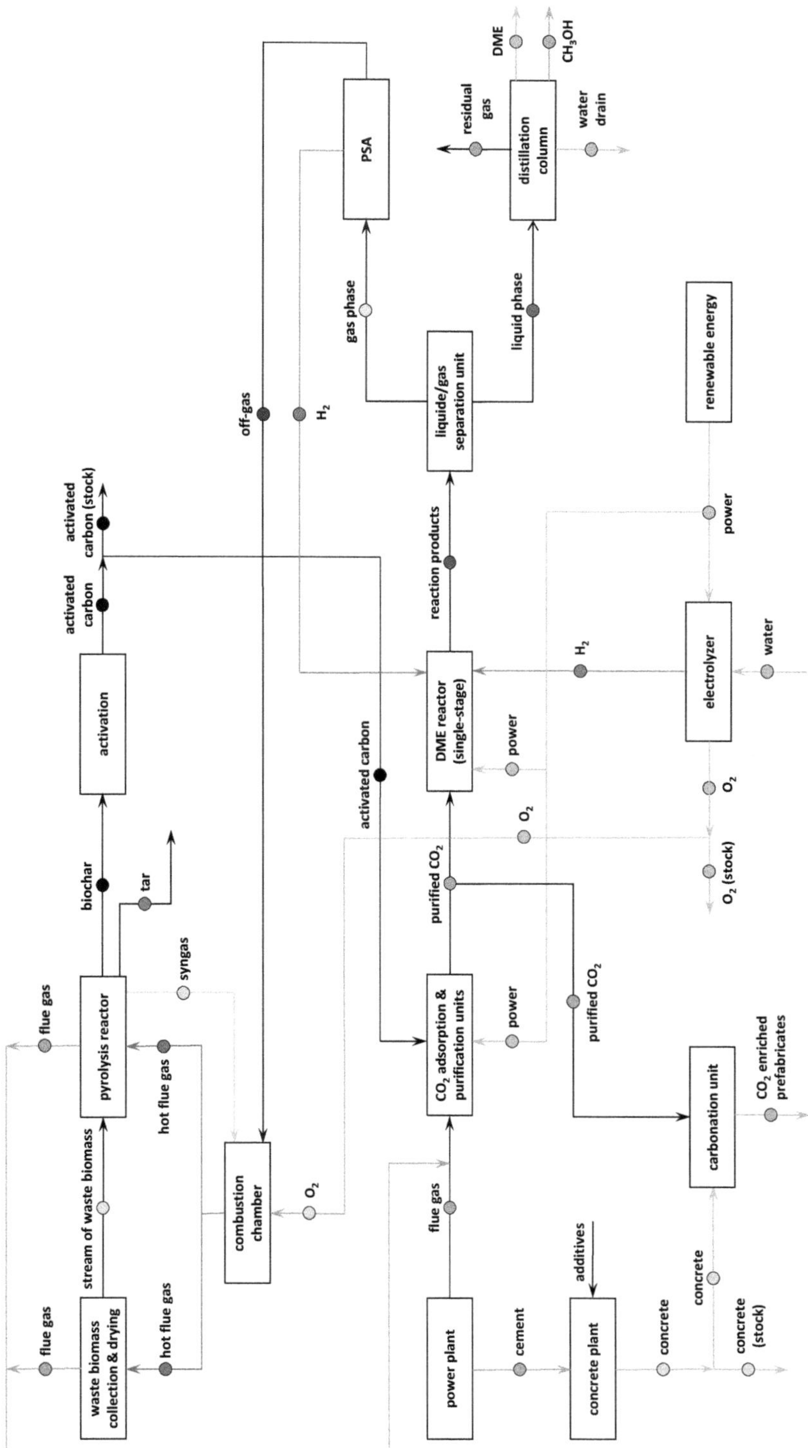

FIGURE 1.4 System of CO_2 bonding in concrete prefabricates and conversion into fuel – a cement plant case study.

being greater than the amount of carbon dioxide in the flue gas filling the closed carbonation chamber, a cyclic injection-release process is required (to release the residual gas which will be replaced with a fresh portion of flue gas). This, however, causes a secondary CO_2 emission with each subsequent cycle of the full carbonation process, due to an increase in its content in the released gas [9]. Therefore, the process analysis comprises the use of enriched CO_2 received from the adsorption installation. In order to carry out the process in an effective manner, it is required to reduce the contents of pollutants, like dust, water vapor, sulfur dioxide, as well as to cool down the gas [10, 11] before it is introduced to the separation unit.

The vacuum-pressure swing adsorption technology has been proposed for CO_2 separation, because there is no need for compressing huge amounts of gas (the product is obtained at a reduced pressure generated by a vacuum pump) and no emission of harmful substances to the atmosphere (as is the case during separation by the absorption method, using e.g.: amines or ammonia solutions). The separation of carbon dioxide is carried out on a solid sorbent, which is activated carbon obtained in the process of waste biomass pyrolysis and activation. This sorbent can also be used in the process of CO_2 purification, if necessary. The obtained high-purity CO_2 reactant is fed together with hydrogen to the DME reactor, where the process of converting the components into DME and methanol takes place. Next, these products, along with by-products, are separated into a liquid phase (where in the distillation column system is separated into: DME, methanol, water drain and residual gas) and gas phase. The latter is separated in the pressure swing adsorption (PSA) unit to recover hydrogen which is recycled to the DME reactor, while the off-gas is directed to the combustion chamber. There, after being mixed with syngas from the pyrolysis unit, it is burned in the oxygen atmosphere to obtain heat to be used to heat up the pyrolysis reactor as well as used for waste biomass drying. This mixture is necessary due to the low calorific value of syngas alone, whose lower heating value (LHV) amounts to about 5.2 MJ/Nm^3 [12]. The generated flue gas is combined with the flue gas coming from a power or cement plant, which causes an increase in CO_2 concentration in the feed to the adsorption unit. The energy necessary to drive the individual units of the system, as well as to produce hydrogen through water electrolysis, comes from renewable energy sources. In addition to manufacturing main products, such as DME, as well as cement prefabricates enriched in carbon dioxide (formed in the concrete carbonation process), the excess of process reactants is obtained, such as activated carbon, oxygen, but also other products, like concrete and methanol that can be sold. The only waste is tar derived from pyrolysis and water drained from the DME reactor, which must be disposed of in accordance with applicable regulations.

1.3 IDENTIFICATION OF CARBON SOURCES

1.3.1 Power plant sources

The increase in electricity demand has made the power generation sector the world's biggest industry which, therefore, has now the largest impact on the environment. Cheap, price stable and widespread coal has become a common and secure source for

TABLE 1.1
Coal-fired power plants flue gas characteristics

		Source			
Parameter	**Unit**	**(a) [16]**	**(b) [18]**	**(c) [19]**	**(d) [20]**
CO_2	vol% [wb]	13.5	10–11	10.8	11.4
O_2	vol% [wb]	11.0	4–5	5.8	6.1
N_2	vol% [wb]	n/a	n/a	Balance	69.4
Ar	vol% [wb]	n/a	n/a	Balance	n/a
H_2O	vol% [wb]	5.5	20–23	9.9	12.6
SO_x	mg/Nm^3	200–400	338[a]–564[a]	<50	~1692[a]
NO_x	mg/Nm^3	250–350	198[a]–330[a]	<160	262[a] (NO) 4[a] (NO_2)
Dust	mg/Nm^3	30	n/a	n/a	n/a
Flue gas temperature	°C	120–145	160–180	55.2	85

[wb] – wet basis.
[a] Calculated from ppm.

energy production [13], while providing high availability of the electricity generation source. Unfortunately, this fuel is the dirtiest of the fossil fuels [13], featuring, among other things, the highest carbon dioxide emission factor, which makes it cumbersome in activities aimed at combating the global warming.

Modern coal-fired boilers differ in many ways e.g., in steam parameters, steam-water circulation, combustion method and design configuration [14], but they generate a similar concentration of carbon dioxide in dry flue gas in the range from 12 to 15 vol% [15]. The parameters for four selected units are presented in Table 1.1.

a. Fluidized-bed boiler, hard coal fired in the Jaworzno II Power Plant in Jaworzno, Poland [16];
b. Forced-circulation-tower type boiler [17], brown coal-fired in the Loy Yang Power in Latrobe Valley, Victoria, Australia [18];
c. Supercritical coal-fired boiler, in the Changchun thermal power plant in Jilin province, China [19];
d. Subcritical lignite coal-fired boiler, in SaskPower's Shand Power Station in Saskatchewan, Canada [20].

1.3.2 Cement Plant Sources

The cement production technology can be divided into four processes [21]: dry process, wet process and its modification, semidry process and semiwet process. Each of them is characterized by a different raw-material preparation and a different configuration of the rotary kiln system [21]. The wet process is more energy intensive than the dry process, due to the need for evaporating the 30% of slurry water before

TABLE 1.2
Cement plant flue gas characteristics

Source					
Parameter	**Unit**	**(e) [23]**	**(f) [23]**	**(g) [23]**	**(h) [24]**
CO_2	vol% [wb]	18	22	29	20[a]
O_2	vol% [wb]	10	7	3	9.5[a]
N_2	vol% [wb]	63	60	54	52[a]
H_2O	vol% [wb]	9	11	13	18.5[a]
SO_x	mg/Nm³	n/a	n/a	n/a	563[a]
NO_x	mg/Nm³	n/a	n/a	n/a	91[a]
Dust	mg/Nm³	10	10	10	n/a
Flue gas temperature	°C	110	130	210	n/a

[wb] – wet basis.
[a] Calculated based on the values of process mass flow.

starting the raw-material calcination process, though it is easier to control the chemistry of the process (and the process is more commonly applied where moist raw feedstocks are available) [22].

The general cement manufacturing structure includes the following stages [22, 23]: pre-heater, pre-calciner, rotary kiln, clinker cooler and final grinding. The characteristics of flue gas, from cement production, depend on the kiln type, the overall plant layout, the amount of air leaking into the system and the mode of operation of the raw mill [23]. Generally, the carbon dioxide concentration in dry flue gas equals 14–35 vol% [23]. About 62% of the total direct CO_2 emission fall to the calcination of carbonate materials, and the rest (approximately 38%), to the combustion of fuels used in the clinker production [21]. The detailed data for selected cement plants are summarized in Table1.2.

e–g. The reference cement kiln [23], taking into account these three assumed cases: (e) interconnected operation with medium air leak (in the amount of 139,800 kg/h), (f) with a low air leak (in the amount of 69,900 kg/h) and (g) for direct operation with no air leak,

h. Capitol Aggregates' Cement Plant in San Antonio, Texas, USA [24].

1.4 EMISSION STANDARDS AND POSSIBILITIES OF FLUE GAS PURIFICATION

Gas purification methods are essential to reducing the concentration of components that are considered undesirable in terms of the environment, process or product. For the removal of pollutants in flue gas or process gas, widely known and applicable methods are used, such as wet, semi-dry or dry (fluidized bed) desulfurization, selective non-catalytic reduction (SNCR) or selective catalytic reduction (SCR) denitration, as well as activated carbon injection (ACI) for mercury removal [25]. The limit values of pollutants emitted in the atmosphere are defined in the standards set by government institutions on the basis of the best available technologies (BAT). Examples of emission standards for selected pollutants in the case of the energy and cement industries are summarized in Tables 1.3 and 1.4. The data relates to new plants: (CN) China, (EU) the European Union, (I) India, (US) the United States.

TABLE 1.3
Emission standards for power plant

Source			**(EU) [27]**			
Pollutant	**Unit**	**(CN) [26]**	**Yearly average**	**Daily average**	**(I) [28]**	**(US) [29]**
SO_2	mg/Nm^3	100[a]	150–200[a,b]	170–220[a,b]	100[a]	1.0 lb/MWh[n]
			80–150[a,c]	135–200[a,c]		
			10–75[a,d]	25–110[a,d]		
			20–75[a,e]	25–110[a,e]		
NO_x	mg/Nm^3	100[a,o]	100–150[a,b,o]	155–200[a,b,o]	100[a]	n/a
			50–100[a,c,o]	80–130[a,c,o]		
			50–85[a,f,o]	80–125[a,f,o]		
			65–85[a,g,o]	80–125[a,g,o]		
PM	mg/Nm^3	30[a]	2–5[a,b]	4–16[a,b]	30[a]	9.0×10^{-2} lb/MWh[n]
			2–5[a,c]	3–15[a,c]		
			2–5[a,h]	3–10[a,h]		
			2–5[a,i]	3–10[a,i]		
Hg	μg/Nm^3	30[a]	<1–3[a,j]	–	30[a]	4.0×10^{-2} lb/GWh[n]
			<1–5[a,k]			
			<1–2[a,l]			
			<1–4[a,m]			
O_2 (reference level)	vol%	6[a]	6[a]		n/a	–

[a] Conditions: 0°C, 101.3 kPa.
[b] <100 MW_{th} total input.
[c] 100–300 MW_{th} total input.
[d] ≥300 MW_{th} total input, PC boiler.
[e] ≥300 MW_{th} total input, fluidized bed boiler.
[f] ≥300 MW_{th} total input, FBC boiler combusting coal and/or lignite and lignite-fired PC boiler.
[g] ≥300 MW_{th} total input, coal-fired PC boiler.
[h] 300–100 MW_{th} total input.
[i] ≥300 MW_{th} total input.
[j] <300 MW_{th}, coal fired.
[k] ≥300 MW_{th}, coal fired.
[l] <300 MW_{th}, lignite fired.
[m] ≥300 MW_{th}, lignite fired.
[n] Gross output, coal-fired unit not low rank as well as low-rank virgin coal.
[o] Calculated to NO_2.

TABLE 1.4
Emission standards for cement plant (kiln flue gas)

Source					
Pollutant	**Unit**	**(CN) [30]**	**(EU) [31]**	**(I) [32]**	**(US) [33, 34]**
SO_2	mg/Nm^3	100[a]	<50–400[a,b,e]	100[a,g] 100[a,h,i] 700[a,h,j] 1000[a,g,k]	0.4 lb/ton clinker[l]
NO_x	mg/Nm^3	320[a,m]	<200–450[a,c,e] 400–800[a,d,e]	600[a,g,h]	1.5 lb/ton clinker[l]
PM	mg/Nm^3	20[a]	<10–20[a,e]	30[a,g,h]	0.02 lb/ton clinker
Hg	mg/Nm^3	0.05[a]	0.05[a,f]	0.05[a,h]	21 lb/MM tons clinker
O_2 (reference level)	vol%	10	10	10	–

[a] Conditions: 0°C, 101.3 kPa.
[b] SO_x expressed as SO_2.
[c] Preheater kilns.
[d] Lepol and long rotary kilns.
[e] Daily average value.
[f] Average over the sampling period (spot measurements for at least half an hour).
[g] Rotary kiln – without coprocessing.
[h] Rotary kiln – with co-processing of wastes.
[i] Pyrolytic sulphur in the limestone is less than 0.25%.
[j] Pyrolytic sulphur in the limestone is between 0.25 and 0.5%.
[k] Pyrolytic sulphur in the limestone is more than 0.5%.
[l] 30-operating day rolling average.
[m] Calculated to NO_2.

For systems designed for carbon dioxide capture from flue gas, some emission limits may not be sufficient in terms of process requirements. In the case of the absorption method, the concentration of sulfur dioxide or nitrogen oxides in the gas, directed to the CO_2 capture installation, should not exceed 10 ppm of SO_2 and 20 ppm of NO_2, because the solvent used gradually degrades as well as corrosive salts and nitric acid are formed [35]. On the other hand, in the adsorption method, where the CO_2 capture process takes place on solid, dry adsorbents, the concentration should not exceed 10 mg/Nm^3 of SO_2 and 10 mg/Nm^3 of NO_x in order to avoid adsorbent degradation [36]. In addition, the gas supplied to the adsorbers (in which CO_2 separation takes place) is dried, which reduces SO_2 and NO_x condensation, allowing the bed to run in a cyclic operation without having to replace or regenerate it at high parameters. Also, catalysts require a reduced concentration of pollutants, which are used, among others, in DME production. As has been pointed out by Bowker [37], a catalyst containing copper is very sensitive to sulfur poisoning.

Therefore, in order to meet the higher purity requirements for flue gas supplied to the CO_2 capture installation and the purity requirements for the product (separated CO_2) to be used, for example, in DME production, additional cleaning may be necessary. This can be done using e.g., the dry method of removing NO_x and SO_2 on adsorbents [38].

As a sorbent, we can use activated carbon produced from waste biomass, but its behavior is different when used for either pure or mixed gases [38]. In a mixture containing SO_2 and NO_x, the adsorption of sulfur dioxide onto activated carbon is promoted [38], while NO is adsorbed quite hardly [39]. It is important because flue gas coming from fossil fuel combustion contains nitrogen oxides (NO_x) which consist of 95% NO and 5% NO_2 [38]. However, activated carbon has also the ability to catalytically oxidize organic and inorganic compounds [40], which can lead to oxidizing NO to NO_2 [41]. Unfortunately, SO_2 present in gases interferes with the sites on the sorbent, which convert NO to NO_2 [42]. Therefore, chemical-impregnated adsorbents are proposed. As demonstrated by Takeuchi et al. [43], impregnation of activated carbon by potassium hydroxide showed good performance in NO_x removal from air. In turn, Guo and Lua [44] confirmed that KOH-impregnated activated carbon had adsorptive capacities comparable to those of some commercial activated carbons. In the case of mercury vapor removal, activated carbons impregnated with potassium iodide, sulfuric acid or sulfur [40] are used.

Purification of flue gas to remove SO_2 and NO_x with the use of solid sorbents – activated carbons impregnated with potassium hydroxide and potassium iodide – was tested during pilot campaigns on real flue gas from a hard coal-fired fluidized-bed boiler in a power plant [10, 36]. The investigation confirms their effectiveness in reducing the flue gas SO_2 concentration from a level of about 155 mg SO_2/Nm^3 down to practically zero, i.e., below 5 mg SO_2/Nm^3 (the overall SO_x removal efficiency amounted to 95.5–100%) [36]. At the same time, the use of this type of sorbent for NO_x absorption did not show a high removal efficiency – the concentration of nitrogen oxides on exit from the adsorber had dropped by only a few mg/Nm^3 [36].

1.5 SYSTEM INPUT PARAMETERS IDENTIFICATION – CASE STUDY

In order to carry out the case study of the carbon chain in industrial carbon dioxide utilization technology, these two CO_2 emission sources were selected:

a. A supercritical circulating fluidized-bed boiler, hard coal-fired, located in the Łagisza Power Plant, owned by TAURON Wytwarzanie S. A., Poland [10, 45];
b. A cement kiln, located in Norway Norcem's cement factory in Brevik, owned by Heidelberg Cement, Norway [46, 47].

The data necessary for the simulation process are summarized in Table 1.5.

TABLE 1.5
Composition and parameters of flue gases – input data for the case study

Parameter	Unit	Power plant (a)	Cement plant (b)
CO_2	mol% [wb]	14.3 [45]	17.8 [46, 47]
O_2	mol% [wb]	7.5 [10]	7.5 [46, 47]
N_2	mol% [wb]	68.3[a]	56.5 [47]
H_2O	mol% [wb]	9.1 [45]	18.2 [46, 47]
Ar	mol% [wb]	0.8[a]	n/a
SO_x	mg/Nm^3	154.6 [10]	0–130 [46]
NO_x	mg/Nm^3	179.8 [10]	180–250 [46]
dust	mg/Nm^3	3.0 [10]	5–10 [46]
Flue gas flow	kg/s	457 [45]	120[a] [47]
Flue gas temperature	°C	123 [45]	165 [47]
Flue gas pressure	kPa	n/a	100 [47]

[wb] – wet basis.
[a] Calculated.

REFERENCES

[1] Friedlingstein, P. et al. *Global Carbon Budget 2020*, Earth System Science Data 2020, 12, 3269–3340, https://doi.org/10.5194/essd-12-3269-2020.

[2] Tans, P.; Keeling, R. Global Monitoring Laboratory Home Page, Mauna Loa CO_2 monthly mean data, NOAA/GML, Scripps Institution of Oceanography. Accessed September 1, 2022, https://gml.noaa.gov/webdata/ccgg/trends/co2/co2_mm_mlo.txt.

[3] Tans, P.; Keeling, R. Global Monitoring Laboratory Home Page, Mauna Loa CO_2 annual mean data, NOAA/GML, Scripps Institution of Oceanography. Accessed September 1, 2021, https://gml.noaa.gov/webdata/ccgg/trends/co2/co2_annmean_mlo.txt.

[4] International Energy Agency Home Page, World Energy Outlook [Online] 2018. Accessed September 10, 2021, https://iea.blob.core.windows.net/assets/77ecf96c-5f4b-4d0d-9d93-d81b938217cb/World_Energy_Outlook_2018.pdf.

[5] Li, Q.; Wu, G.; Johnston, C. M.; Zelenay, P. Direct dimethyl ether fuel cell with much improved performance. *Electrocatalysis* **2014**, *5*, 310–317, https://doi.org/10.1007/s12678-014-0196-z.

[6] Wawrzyńczak, D. DME – alternative green fuel. In *Modern technologies for energy conversion and management*; Majchrzak-Kucęba, I., Mirek, P., Bień, J. Eds.; The Publishing Office of Czestochowa University of Technology, Częstochowa, 2019, 96–107.

[7] Shao, Y.; Zhou, X.; Monkman S. *A new CO_2 sequestration process via concrete products production.* In 2006 IEEE EIC Climate Change Conference, 1–6, 2006, https://doi.org/10.1109/EICCCC.2006.277189.

[8] Abbas, Z.; Mezher, T.; Abu-Zahra, M. R. M.; Evaluation of CO_2 purification requirements and the selection of processes for impurities deep removal from CO_2 product stream. *Energy Procedia* **2013**, *37*, 2389–2396, https://doi.org/10.1016/j.egypro.2013.06.120.

[9] He, Z.; Wang, S.; Mahoutian, M.; Shao, Y.; Flue gas carbonation of cement-based building products. *Journal of CO_2 Utilization* **2020**, *37*, 309–319, https://doi.org/10.1016/j.jcou.2020.01.001.

[10] Wawrzyńczak, D.; Majchrzak-Kucęba, I.; Srokosz, K.; Kozak, M.; Nowak, W.; Zdeb, J.; Smółka, W.; Zajchowski, A. The pilot dual-reflux vacuum pressure swing adsorption unit for CO_2 capture from flue gas. *Separation and Purification Technology* **2019**, *209*, 560–570, https://doi.org/10.1016/j.seppur.2018.07.079.

[11] Majchrzak-Kucęba, I.; Wawrzyńczak, D.; Ściubidło, A.; Zdeb, J.; Smółka, W.; Zajchowski, A. Stability and regenerability of activated carbon used for CO_2 removal in pilot DR-VPSA unit in real power plant conditions. *Journal of CO_2 Utilization* **2019**, *29*, 1–11, https://doi.org/10.1016/j.jcou.2018.11.003.

[12] Elsner, W.; Wysocki, M.; Niegodajew, P.; Borecki, R. Experimental and economic study of small-scale CHP installation equipped with downdraft gasifier and internal combustion engine. *Applied Energy* **2017**, *202*, 213–227, https://doi.org/10.1016/j.apenergy.2017.05.148.

[13] Breeze, P. *Coal-fired power plants. Power generation technologies.* 2nd ed.; Newnes, 2014, 29–65, https://doi.org/10.1016/B978-0-08-098330-1.00003-X.

[14] Zhang, K.; Zhang, Y.; Guan, Y.; Zhang, D. Boiler design for ultra-supercritical coal power plants. In *Ultra-supercritical coal power plants*; Zhang, D., Ed.; Woodhead Publishing, 2013, 104–130, https://doi.org/10.1533/9780857097514.1.104.

[15] Gale, J.; Bradshaw, J.; Chen, Z.; Garg, A.; Gomez, D.; Rogner, H.-H.; Simbeck, D.; Williams, R.; Toth, F.; Vuuren, D. Sources of CO_2. In *IPCC special report on carbon dioxide capture and storage*; Metz, B., Davidson, O., Coninck, H., Loos, M., Meyer, L., Eds.; Cambridge University Press, 2015, 77–103. Accessed September 10, 2021, https://www.ipcc.ch/site/assets/uploads/2018/03/srccs_chapter2-1.pdf.

[16] Stec, M.; Tatarczuk, A.; Więcław-Solny, L.; Krótki, A.; Ściążko, M.; Tokarski, S. Pilot plant results for advanced CO_2 capture process using amine scrubbing at the Jaworzno II Power Plant in Poland. *Fuel* **2015**, *151*, 50–56, https://doi.org/10.1016/j.fuel.2015.01.014.

[17] Bertoli, R.; Dyson, G.; Engineering analysis lengthens pump's lifespan. *World Pumps* **2018**, *6*, 30–32, https://doi.org/10.1016/S0262-1762(19)30111-7.

[18] Artanto, Y.; Jansen, J.; Pearson, P.; Do, T.; Cottrell, A.; Meuleman, E.; Feron, P. Performance of MEA and amine-blends in the CSIRO PCC pilot plant at Loy Yang Power in Australia. *Fuel* **2012**, *101*, 264–275, https://doi.org/10.1016/j.fuel.2012.02.023.

[19] Feron, P.; Conway, W.; Puxty, G.; Wardhaugh, L.; Green, P.; Maher, D.; Fernandes, D.; Cousins, A.; Shiwang, G.; Lianbo, L.; Hongwei, N.; Hang, S. Amine based post-combustion capture technology advancement for application in Chinese coal fired power stations. *Energy Procedia* **2014**, *63*, 1399–1406, https://doi.org/10.1016/j.egypro.2014.11.149.

[20] Giannaris, S.; Bruce, C.; Jacobs, B.; Srisang, W.; Janowczyk, D. Implementing a second generation CCS facility on a coal fired power station – results of a feasibility study to retrofit SaskPower's Shand power station with CCS. *Greenhouse Gases: Science and Technology* **2020**, *10*, 506–518, https://doi.org/10.1002/ghg.1989.

[21] John, P. J. Parametric studies of cement production processes. *Journal of Energy* **2020**, *2020*, https://doi.org/10.1155/2020/4289043.

[22] Bosoaga, A.; Masek, O.; Oakey, J. E. CO_2 capture technologies for cement industry. *Energy Procedia* **2009**, *1*, 133–140, https://doi.org/10.1016/j.egypro.2009.01.020.

[23] Anantharaman, R.; Berstad, D.; Cinti, G. et al. *CEMCAP framework for comparative techno-economic analysis of CO_2 capture from cement plants*; D3.2, SINTEF-ER, 2017. Accessed September 1, 2021, https://www.sintef.no/globalassets/-sintef-energi/cemcap/d3.2-cemcap-framework-for-comparative-techno-economic-analysis-of-co2-capture-from-cement-plants_.pdf.

[24] Jones, J.; Barton, C.; Clayton, M.; Yablonsky, A.; Lergere, D. *SkyMine carbon mineralization pilot project*; Final phase 1 topical report, United States, September 2010, https://doi.org/10.2172/1027801.

[25] Yan, Y.; Mao, Z.; Luo, J.; Du, R.; Lin, J. Simultaneous removal of SO_2, NOx and Hg^0 by O_3 oxidation integrated with bio-charcoal adsorption. *Journal of Fuel Chemistry and Technology* **2020**, *48*(12), 1452–1460, https://doi.org/10.1016/S1872-5813(20)30092-X.
[26] Emission standard of air pollutants for thermal power plants, (GB 13223-2003). Accessed September 1, 2021, https://english.mee.gov.cn/Resources/standards/Air_Environment/Emission_standard1/201201/W020110923324406748154.pdf.
[27] Comission implementing decission (EU) 2017/1442 of 31 July 2017 establishing best available techniques (BAT) conclusions, under Directive 2010/75/EU of the European Parliament and of the Council, for large combustion plants. Accessed September 1, 2021, https://eur-lex.europa.eu/legal-content/EN/TXT/PDF/?uri=CELEX:32017D1442.
[28] Environment standards for thermal power plants along with the corrigendum, S.O.3305(E). Accessed September 1, 2021, https://moef.gov.in/wp-content/uploads/2017/08/Thermal_plant_gazette_scan.pdf.
[29] National Emission Standards for Hazardous Air Pollutants for Source Categories (Part 63), National Emission Standards for Hazardous Air Pollutants: Coal- and Oil-Fired Electric Utility Steam Generating Units (Subpart UUUUU). Accessed September 1, 2021, https://www.ecfr.gov/current/title-40/chapter-I/subchapter-C/part-63/subpart-UUUUU.
[30] Emission standard of air pollutants for cement industry (GB 4915-2013). Accessed September 1, 2021, https://english.mee.gov.cn/Resources/standards/Air_Environment/Emission_standard1/201605/W020160511519363364177.pdf.
[31] Commission implementing Decision (2013/163/EU) of 26 March 2013 establishing the best available techniques (BAT) conclusions under Directive 2010/75/EU of the European Parliament and of the Council on industrial emissions for the production of cement, lime and magnesium oxide. Accessed September 1, 2021, https://eur-lex.europa.eu/legal-content/EN/TXT/PDF/?uri=CELEX:32013D0163.
[32] Environment Standards for Cement Plant, G.S.R.497(E). Accessed September 1, 2021, https://moef.gov.in/wp-content/uploads/2017/08/Cement_Co_processing_Gazette.pdf.
[33] National Emission Standards for Hazardous Air Pollutants for Source Categories (Part 63), National Emission Standards for Hazardous Air Pollutants from the Portland Cement Manufacturing Industry (Subpart LLL). Accessed September 1, 2021, https://www.ecfr.gov/current/title-40/chapter-I/subchapter-C/part-63/subpart-LLL.
[34] Standards of Performance for New Stationary Sources (Part 60), Standards of Performance for Portland Cement Plants (Subpart F), Standards (§ 60.62). Accessed September 1, 2021, https://www.ecfr.gov/current/title-40/chapter-I/subchapter-C/part-60/subpart-F/section-60.62.
[35] Plaza, M. G.; Martínez, S.; Rubiera, F. CO_2 capture, use, and storage in the cement industry: State of the art and expectation. *Energies* **2020**, *12*, 5692, https://doi.org/10.3390/en13215692.
[36] Majchrzak-Kucęba, I.; Wawrzyńczak, D.; Zdeb, J.; Smółka, W.; Zajchowski; A. Treatment of flue gas in a CO_2 capture pilot plant for a commercial CFB boiler. *Energies* **2021**, *14*, 2458, https://doi.org/10.3390/en14092458.
[37] Bowker, M. Methanol synthesis from CO_2 hydrogenation, *ChemCatChem* **2019**, *11*, 4238–4246, https://doi.org/10.1002/cctc.201900401.
[38] Gholami, F.; Tomas, M.; Gholami, Z.; Vakili, M. Technologies for the nitrogen oxides reduction from flue gas: a review. *Science of The Total Environment* **2020**, *714*, 136712, https://doi.org/10.1016/j.scitotenv.2020.136712.
[39] Takeuchi, Y.; Yanagisawa, K.; Tanaka, Y.; Tsuruoka, N. Removal of nitrogen oxides from air by chemicals-impregnated carbons. *Korean Journal of Chemical Engineering* **1997**, *14*, 377–381, https://doi.org/10.1007/BF02707055.
[40] Henning, K.-D.; Schäfer, S. Impregnated activated carbon for environmental protection. *Gas Separation & Purification* **1993**, *7*(4), 235–240, https://doi.org/10.1016/0950-4214(93)80023-P.

[41] Wang, Z.; Kuang, H.; Zhang, J.; Chu, L.; Ji, Y. Nitrogen oxide removal by coal-based activated carbon for a marine diesel engine. *Applied Science* **2019**, *9*, 1656, https://doi.org/10.3390/app9081656.
[42] Stencel, J. M.; Rubel, A. M. Coal-based activated carbons: NO_x and SO_2 postcombustion emission control. *Coal Science and Technology* **1995**, *24*, 1791–1794, https://doi.org/10.1016/S0167-9449(06)80163-1.
[43] Takeuchi, Y.; Yanagisawa, K.; Tanaka, Y.; Noriyuki T. Removal of nitrogen oxides from air by chemicals-impregnated carbons. *Korean Journal of Chemical Engineering* **1997**, *14*, 377–381, https://doi.org/10.1007/BF02707055.
[44] Guo, J.; Lua, A. C. Adsorption of sulphur dioxide onto activated carbon prepared from oil-palm shells with and without pre-impregnation. *Separation and Purification Technology* **2003**, *30*(3), 265–273, https://doi.org/10.1016/S1383-5866(02)00166-1.
[45] Myöhänen, K.; Hyppänen, T.; Pikkarainen, T.; Eriksson, T.; Hotta, A. Near zero CO_2 emissions in coal firing with oxy-fuel circulating fluidized bed boiler. *Chemical Engineering & Technology* **2009**, *32*, 355–363, https://doi.org/10.1002/ceat.200800566.
[46] Knudsen, J. N.; Bade, O. M.; Askestad, I.; Gorset, O.; Mejdell, T. Pilot plant demonstration of CO_2 capture from cement plant with advanced amine technology. *Energy Procedia* **2014**, *63*, 6464–6475, https://doi.org/10.1016/j.egypro.2014.11.682.
[47] Schakel, W.; Hung, C. R.; Tokheim, L. A.; Strømman, A. H.; Worrell, E.; Ramírez, A. Impact of fuel selection on the environmental performance of post-combustion calcium looping applied to a cement plant. *Applied Energy* **2018**, *210*, 75–87, https://doi.org/10.1016/j.apenergy.2017.10.123.

2 CO_2 *A useful reactant*

Izabela Majchrzak-Kucęba

CONTENTS

2.1 INTRODUCTION

Carbon dioxide is a colorless, odorless, non-flammable gas slightly acidic in taste, which is generated on a large industrial scale mainly as a by-product in the production of ammonia and hydrogen. The majority of thus produced carbon dioxide is of a high final purity of up to 99.9%, if an appropriate cleaning process is used. Production of high-purity carbon dioxide is extremely important, as a considerable part of it is used in the food and beverage industry, where the main criteria for carbon dioxide quality include the absence of odor and taste [1]. Impurities, such as sulfur compounds, oils and hydrocarbons, which may affect these criteria, must be removed from the CO_2 stream. Carbon dioxide is sold in either liquid or solid form. A large part (approx. 50%) of produced carbon dioxide is utilized on-site at the urea and methanol production facility. In those applications, carbon dioxide is used in a gaseous form. Another important application of CO_2 in a gaseous form is for Enhanced Oil Recovery (EOR). The remaining part of the carbon dioxide is used in a liquid or solid form, as the transport of gaseous carbon dioxide is not economically viable.

DOI: 10.1201/9781003336587-2

Liquid carbon dioxide is used, e.g., for promoting plant growth or as a refrigerant in the food industry [1]. Although carbon dioxide is commonly regarded as the main factor contributing to climate change, it can also be perceived as a potential source of carbon in chemical reactions. It can make a useful reactant – chemical substrate which, because of its abundance and availability, has the potential to partially substitute for fossil raw materials used in the chemical production of compounds, such as urea, polymers, etc. [1]. In view of the physicochemical properties of CO_2 and a wide spectrum of its potential applications, it becomes an increasingly interesting idea to utilize considerable amounts of CO_2 waste streams originating from the power industry and other industries, which, after appropriate processing, could provide a useful source of carbon in many processes.

2.2 INDUSTRIAL SOURCES OF CO_2

The anthropogenic emission of CO_2 from industrial activity occurs not only as a result of fuel combustion but also the oxidation of carbonaceous reducing agents and the release of CO_2 as an impurity from manufacturing processes. Even though the main emphasis in the implementation of CO_2 emission reduction technologies is being currently placed on the energy sector (fossil-fuels-based power), which is responsible for approx. 71% of the world's emissions of CO_2 from large stationary sources (>0.1 Mt CO_2/year), more and more attention is being given to the problem of reducing CO_2 emissions also from other branches of industry, such as the cement, petrochemical or metallurgical industries. These industries emit the remaining 29% of CO_2, of which cement plants account for ~7%, refineries for ~6%, and the metallurgical industry accounts for ~5% (Figure 2.1a) [2]. The average CO_2 emissions by source are also the highest for the energy sector and amounts to 5.72 Mt CO_2/source (Figure 2.1b) [2].

Combustion of fossil fuels is the largest source of CO_2 emissions, as it is used in various energy applications, such as energy generation, petroleum refining or industrial activity. CO_2 comes also from chemical and petrochemical processes, such as production of ammonia or hydrogen, the cement industry and the iron and

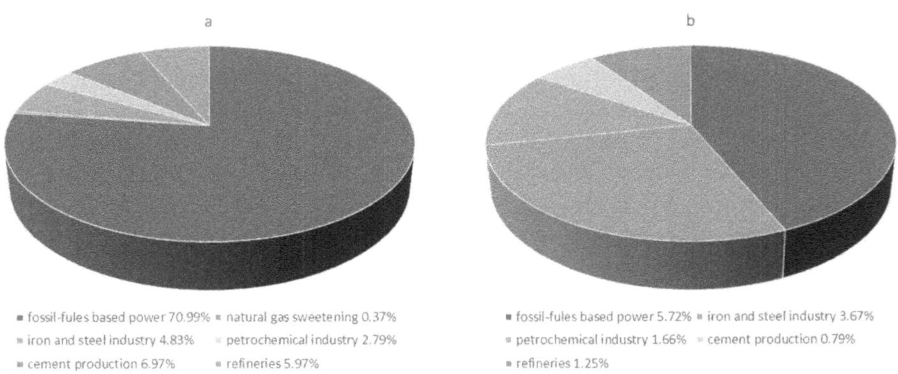

FIGURE 2.1 The CO_2 emissions from the global large stationary sources (>0.1 Mt CO_2/year): (a) % of total CO_2 emissions, (b) average CO_2 emissions by source (Mt CO_2/source) [6].

TABLE 2.1
Flue gases originating from different industrial-energy sectors

CO_2 source	CO_2 concentration in flue gas (s% by volume)	Chemical species (impurities) in flue gas
	Fossil-fuels based power	
Coal boiler	12–14 [6] 14 [4] 12–15 [2] 12–15 [3]	NO_x, SO_x, N_2, H_2O, O_2, Hg/As, HCl, HF, particulates [5]
Natural gas boiler	7–10 [2] 3–10 [3, 7]	NO_x, SO_2, H_2O, O_2 [3, 7]
Natural gas turbine	3 [2] 3–4 [4] 3–4 [3]	NO_x, SO_x, N_2, H_2O, O_2, CO [5]
	Petrochemical industry	
Ammonia production	~100 [2, 7] >98 [3]	H_2, H_2O, N_2, CH_4 [7, 3]
Ethylene production	12 [2] 8 [6]	–
Iron and steel industry	3–27 [4] 17–35 [7, 3] 20–27 [6]	NO_x, SO_x, N_2, O_2 particulates, HCl, H_2O, hydrocarbons, metals, Hg^{2+} [5]; NO_x, SO_x, BTEX (benzene, toluene, ethylbenzene, xylene), PHAs (polycyclic aromatic hydrocarbons) [7, 3]
	Natural gas sweetening	
Refineries	3–13 [2, 4, 6]	–
Cement industry	20 [2] 25[4] 14–33 [7] 14–33 [6]	NO_x, SO_x, NH_3, HCl, HF, VOCs (volatile organic compounds) [7, 3]; CO, NO_x, SO_x, HCl, acetone, benzene, toluene, chloromethane [8]
Lime production	47 [8]	CO_2, CO, SO_2, NO_x, As, Cd, Cr, Ni, HCl [8]; H_2O, HCl, NH_4, Ca, Mg, K, Na [8]

steel metallurgy industry. The characteristic of flue gases originating from different industrial-energy sectors is given in Table 2.1. The concentration of CO_2 in combustion gas depends, among other things, on the type of fuel and the conditions of combustion (i.e., the excess air level). When using gas turbines, we have generally a low CO_2 flue gas content of approx. 3–4 vol%, compared to coal-fired boilers which emit flue gas of CO_2 of 12–15 vol%. Unfortunately, high CO_2–concentration streams coming from petrochemical production and natural gas sweetening have a small share in the total CO_2 emission (for ~3%). This is important inasmuch as carbon dioxide is much easier and less costly to separate from streams of high CO_2 concentration than from those of low CO_2 concentration [2].

As shown in Table 2.1, CO_2 can be captured from many major industrial emission sources because of, inter alia, its concentration in the exhaust gas stream [2–9]. The implementation of the CO_2 capture and utilization technology for the majority of industrial activities requires, however, the stage of capturing appropriate to flue gases of low CO_2 concentration. In some instances, an industrial activity already uses a certain form of CO_2 removal or capture as an integral part of the process and therefore emits a relatively pure CO_2 stream. An example can be the processing of natural gas and generation of hydrogen for the production of ammonia and subsequent production of fertilizers. Table 2.1 also presents the content of impurities in flue gas, significant in the context of CO_2 separation from flue gas and its further disposal.

2.3 POSSIBILITIES OF INDUSTRIAL CO_2 CAPTURE

The climate change issues and activities aimed at reducing the anthropogenic CO_2 emissions have contributed to the development of the CCS and CCU (Carbon Capture and Storage and Carbon Capture and Utilization) technologies. The interest in these technologies results from the need for the reduction of huge amounts of CO_2 waste streams originating chiefly from power plants and other energy-intensive industries [2–9]. The high potential of the CCS/CCU technology in the energy-intensive branches of industry results not only from the considerable CO_2 emissions, but also from the fact that many industrial processes generate flue gases of high CO_2 concentration, which significantly reduces the costs of the CCS/CCU technology. The starting process, both in the CCS and CCU technology, is the CO_2 capture process. CO_2 capture technologies are divided into pre-combustion or post-combustion and oxy-combustion systems [10–11]. Pre-combustion is a process in which carbon dioxide is removed prior to the combustion process. This process is carried out in the case of coal gasification and gas and oil fuel reforming processes that convert carbon compounds into fuels, whose main components are CO and H_2. These components are obtained as a result of the reaction of a fuel with deficient air or with water vapor. Carbon monoxide reacts with water vapor in the catalytic reactor, whereby CO_2 and hydrogen form. The carbon dioxide can be removed, while retaining a hydrogen-rich fuel [10]. Post-combustion is a process in which CO_2 is captured from combustion gas. The course of the capture process does not affect the fuel combustion processes [10]. Oxy-combustion is an oxygen–fuel combustion process that uses for combustion of pure oxygen instead of air (or air considerably enriched by prior removal of the nitrogen). To reduce the furnace temperature and increase the CO_2 concentration, this technology uses the recirculation of portions of the combustion gas, composed mainly of CO_2 and O_2. In oxy-combustion, as compared to air combustion, a gas stream of a much smaller volume is obtained, which is composed primarily of CO_2 in high concentration, water vapor, N_2 (left from the air separation process) and O_2 (deriving from the excess of the oxidant). This combustion gas composition makes the CO_2 separation from this stream easier and less energy intensive, compared to the separation from combustion gas after the conventional air combustion process [10]. Pre-combustion, post-combustion and oxy-combustion systems are aimed at increasing the CO_2 concentration in gases to be separated. The main technologies of flue gas CO_2 separation include absorption, adsorption, membranes and cryogenics (Figure 2.2) [11].

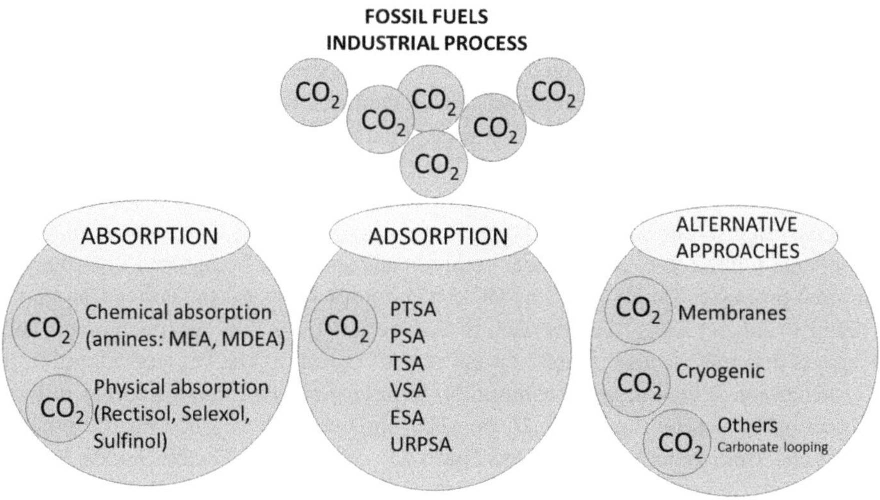

FIGURE 2.2 Schematic diagram of CO_2 capture technologies.

The main consideration that influences the selection of a CO_2 separation method is the concentration and pressure of CO_2 in the combustion gas. The most technologically mature separation technology is considered to be chemical absorption technology using amine solutions. The main advantage of chemical absorption is the fact that it enables the removal of 75–96% CO_2 under industrial conditions, while yielding a gas stream of very high CO_2 concentration (>99%), which considerably facilitates its subsequent storage and/or utilization [11]. With the intensive development of the absorption technique of CO_2 capture from combustion gas, which has been observed for many years now, an adsorption CO_2 capture technique relying on solid adsorbent is arousing an increasing interest [12–19]. The PSA/VPSA adsorption method is a well-established combustion gas CO_2 separation process owing to the ease of its application and relatively wide temperature and pressure ranges. The main advantages of adsorption methods include low energy consumption, simple operation, easy maintenance and flexibility in design to meet different requirements of demand, solid adsorbents without the need for their periodical replenishment, no toxic components in environmental emissions, flexibility of the plant's operation, and the process run cyclically [12–19].

2.4 CARBON DIOXIDE APPLICATION – STORAGE AND UTILIZATION OPTIONS

Until recently, the focus of the proposed CO_2 capture technologies was primarily on capturing CO_2 of the highest purity with the highest possible degree of recovery. The primary proposed option of the disposal of captured CO_2 is the concept of its underground storage on water-bearing saline levels and injecting it to geological formations, such as depleted hydrocarbon deposits. This method has been used for

EOR and also Enhanced Gas Recovery (EGR) [3]. Despite numerous studies on this subject and high potential in terms of storage locations, unfortunately, this method has still not been socially accepted. In recent years, more and more efforts have been dedicated to technologies whereby waste CO_2 captured from power stations or other branches of industry could become a source of carbon for production of fuels, chemicals and various types of materials [3]. The CCU (Carbon Capture and Utilization) technology being developed assumes that after having been separated and recovered from coal-fired power stations and other branches of industry, CO_2 can be subjected to conversion into a fuel or chemical compounds capable of being used in numerous chemical processes [3, 20–30]. The CCU technology relies on the conviction that it is better to subject a gas to conversion than store it (CCS). This approach opens new prospects and may arouse interest on the part of industry. The key consideration in CO_2 utilization is the energy consumption. In order to ensure that a selected utilization technology emits the least CO_2 possible, the energy necessary for the process must come from renewable sources. That CO_2 constitutes valuable goods is confirmed by the fact that about 230 Mt of carbon dioxide is used each year in the world. The largest purchaser of CO_2 is the fertilizer industry that consumes about 130 Mt CO_2 (57% of the total use) for urea production, followed by the oil and gas industry consuming 70 to 80 Mt CO_2 chiefly for EOR (34% of the total use). Other commercial uses of CO_2 include production of food (3%) and beverages (3%), production of metals (2%), greenhouse plant growth stimulation and others (4%) [3, 20]. Currently, the majority of commercial applications involve the direct use of CO_2 (without conversion). The direct use of CO_2 is applicable in many industries and sectors where CO_2 is used as a refrigerant and as a beverage carbonation or food preservation agent. The use of CO_2 in the food and beverage industry amounts to around 11 Mt/year. EOR and Enhanced Coal Bed Methane (ECBM) recovery are also pathways of direct CO_2 utilization. Whereas the ECBM technique is still not economically viable, EOR has been used in the petroleum industry for 40 years represents one of the main markets for CO_2 utilization [20]. The new paths of CO_2 utilization comprise conversion into fuels, chemicals and building materials. The production of fuels and chemicals based on CO_2 is energy-intensive and requires large amounts of hydrogen. As a source of carbon, CO_2 enables the conversion of hydrogen into a fuel which is easier in use than, e.g., aviation fuel. CO_2 may also substitute for fossil fuels as the source in the production of numerous chemicals and polymers. A less energy-intensive way of CO_2 utilization is production of building materials relying on the reaction of CO_2 with minerals or waste materials, such as slag or fly ash, to form carbonates. Figure 2.3 illustrates the main directions in CO_2 utilization [20].

Among many various possibilities of CO_2 utilization, according to [3], for CO_2 streams captured from the power industry and other industries, the most perspective seems to be the following: carbonatization of minerals for building materials production and chemical or biological conversion of carbon dioxide into fuels and chemicals. The mineral carbonization of minerals comprises the reaction of (primarily calcium or magnesium silicates) with CO_2 resulting in inert carbonates. In addition to minerals, also waste materials from steel production or from the cement industry, which are rich in calcium and magnesium oxides, can be used for the production of carbonates in the presence of CO_2 [2–3, 24–27]. Mineral carbonates are used

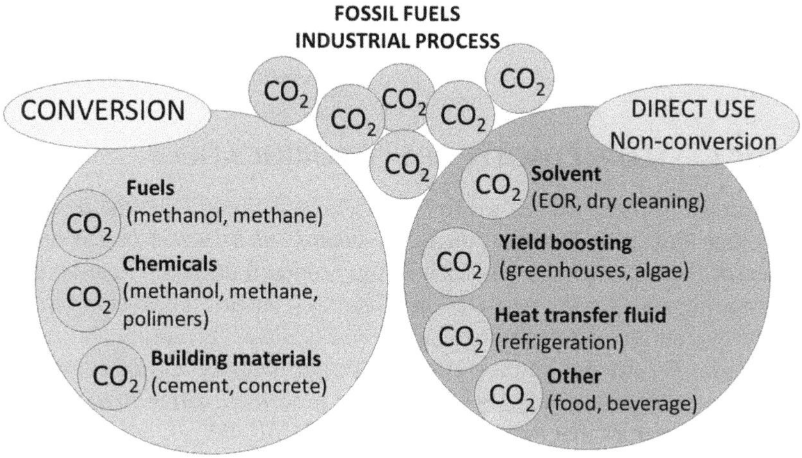

FIGURE 2.3 Schematic diagram of CO_2 utilization technologies.

in the building industry as carbonate blocks instead of concrete blocks produced based on Portland cement (PC) due to their carbon footprint. Production of fuels and chemicals [2] represents a future and high potential for CO_2 utilization, estimated at 500 Mt/year. Captured CO_2 may be used for the production of various chemicals, such as urea, formic acid, salicylic acid, organic carbonates, and polycarbonates. Urea is considered the main product and is used chiefly as an agricultural fertilizer, but also in the production of some pharmaceuticals or in the synthesis of polymers. Carbon dioxide can also be transformed into fuels, such as methane, methanol and synthesis gas [2–3]. Reforming and hydrogenation of dry methane are considered the principal pathway of CO_2 conversion into fuels. The dry reforming of methane is an endothermic reaction, where CO_2 is used in place of steam to react with methane, thus generating synthesis gas. The generated synthesis gas can also be utilized via Fisher-Tropsch (FT) reaction to produce various fluid fuels. The conventional methanol production process is based on the conversion of synthesis gas which is normally generated from natural gas. Methanol can be used as an energy carrier in the transport sector and as a raw-material and solvent for the production of other chemicals (e.g., acetic acid, formaldehyde, methylamine) or as an additive to fuels [2–3, 21–23, 25–26, 28–29]. Biological conversion of CO_2 is another method of CO_2 utilization with a significant sequestration potential. It comprises the absorption and bonding of CO_2 by algae and other land crops in the process of photosynthesis. Microalgae are a type of microscopic algae that function in aquatic environments and are used in various applications from the production of biofuels to animal fodders. In the photosynthesis process, microalgae make use of light to convert CO_2 to organic carbon needed for the production of cell compounds/substances. Microalgae are a prospective means of CO_2 utilization because of their fast growth rate in the presence of high CO_2 concentrations, which enables the bonding of CO_2 to occur 50 times faster than for land-grown plants. A path of biological CO_2 conversion is enriching carbon dioxide in greenhouses, thereby raising CO_2 concentration in greenhouse beds to increase the crop yields [3]. Studies carried out on this subject have shown that CO_2 levels of up to 1200 ppm are advantageous to most of agricultural crops and improve the yields by as much as 30% and lower the levels of evapotranspiration of crops [3, 30].

2.5 SUGGESTED METHODS OF CO_2 PURIFICATION

The assessment of impurities in waste CO_2 streams captured from the power industry and other industries is important in the context of selection of the method of utilization of the captured CO_2, as each of the proposed utilization methods has its specific requirements for the purity of the waste CO_2 stream. The level of impurities in a waste CO_2 stream is largely influenced by, e.g., the CO_2 capture and separation method used. CO_2 streams coming from the power industry and other branches of industry, due to the different types of occurring processes, differ also in the ranges and levels of impurities that they contain (Table 2.1) [31–41].

2.5.1 Impurities Contained in a CO_2 Stream Captured from a CCS Power Plant

Impurities may form in many different ways (Figure 2.4). They originate from (1) the fuel and the combustion process used, (2) from the employed CO_2 capture method and from (3) the transport of the captured CO_2. The composition of a waste CO_2 stream depends, inter alia, on the mode of power plant operation and the technologies used for CO_2 separation and impurity removal. Impurities contained in CO_2 streams coming from different CO_2 capture technologies can be generally divided into three main categories resulting from: fuel oxidation, excessive oxidation/deaeration and process liquids [31].

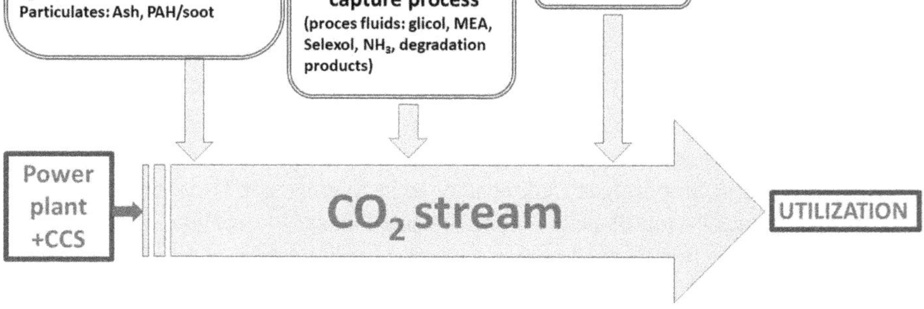

FIGURE 2.4 Impurities in a CO_2 stream from a CCS power plant.

Water is the main combustion product and is regarded as an impurity in the CO_2 stream. Coal present in the fuel contains sulfur, chlorine and mercury, which are released after either complete or incomplete burning and form compounds in a gaseous phase, which may remain, to some degree, as impurities in the CO_2 after its capture and compression. The oxidizer in the combustion process, namely air, also supplies residual impurities, such as N_2, O_2 and Ar. The same impurities may also form as a result of air entering the process. Absorption liquids used in the CO_2 separation process, such as MEA (Monoethanoloamine) for capture after the combustion process or Selexol used in capture before the combustion process, as well as the products of their degradation, may also be transferred to the CO_2 stream, thus constituting another group of impurities [31–34]. The products of oxidation of coal and/or biomass generate a wider range and higher level of CO_2 impurities, compared to CO_2 derived from the CCS combustion of natural gas. The major and minor products of the complete oxidation of coal and biomass form common impurities, namely: water, SO_x, NO_x and halogens. Partial oxidation products, such as carbon monoxide (CO) and hydrogen sulfide (H_2S), may form in fuel-rich conditions that are found in gasifiers used in Integrated Gasification Combined Cycles (IGCC). Volatile substances containing hydrogen and light hydrocarbons form as the result of fuel degassing. Biomass fuels have a higher level of alkaline metals, as compared to bituminous coal, and can form a class of CO_2 impurities, the main species of which are chlorides, sulfates and potassium and sodium hydroxides. Trace metals contained in the fuel may be released to the gaseous phase during combustion and spread within the CO_2 stream. Those metals can occur in the CO_2 stream in either an elementary or oxidized form, such as mercury dichloride ($HgCl_2$), and may require removal for operational and environmental reasons. Particulate solids in the form of ash and soot with the precursors of polycyclic aromatic hydrocarbons (PAH) are another type of oxidizing impurities. Contamination of the CO_2 stream with oxygen, nitrogen and argon may occur due to the excess of the combustion oxidizer or air entering the boiler. Impurities may also originate from a power plant or a CCS process – as machine grease or metals – but are not present in CO_2 streams at levels that would give rise to a concern. The range and level of impurities coming from three currently available types of CO_2 capture technology, namely: pre-combustion capture; post-combustion capture; and oxy-combustion capture, are different [31–34]. Figure 2.5 illustrates the typical composition of combustion gas and that of captured CO_2, depending on the employed capture technology, namely, post-combustion, pre-combustion and oxy-combustion, respectively [10].

Figure 2.5 summarizes the main impurities coming from different CO_2 capture technologies. The highest purity, ~99.6%, is shown by the CO_2 stream from post-combustion capture, followed by the CO_2 stream from pre-combustion capture, ~98%, and then oxy-combustion capture, ~96%. The highest-concentration impurities in the oxy-combustion stream are O_2, N_2 and Ar, but also SO_x and Hg can occur at a certain level and may be the cause of concern due to potential corrosion and toxicity. The main impurities coming from pre-combustion CO_2 capture are: H_2S – unfavorable because of its corrosion potential – and H_2, which may be present in the CO_2 stream in a considerable concentration, which leads to a relative increase in transport cost. In post-combustion capture, the levels of impurities are

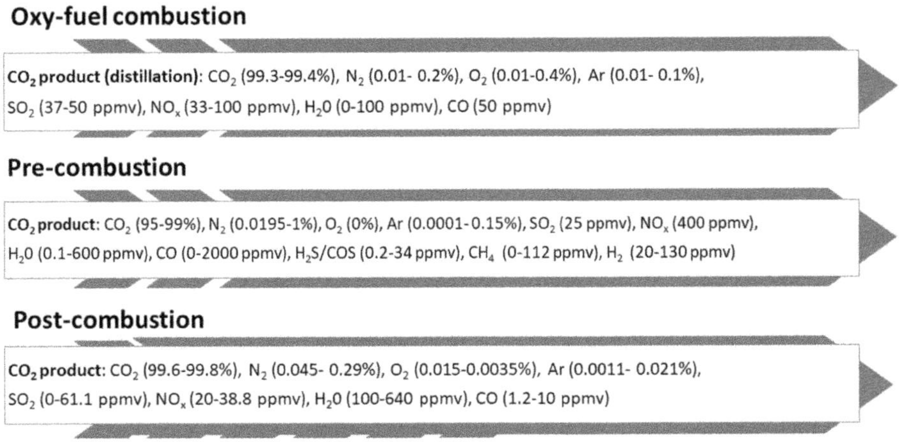

FIGURE 2.5 CO_2 impurities in different capture and purification technologies [31].

lower than in standard capture technologies, and N_2, H_2O and O_2 are regarded as the main impurities in the highest concentration. To sum up, the type and presence of impurities in oxy-combustion, pre-combustion and post-combustion may considerably differ. Non-condensing components, i.e., N_2, Ar, O_2, CO, N_2 and water, are common impurities in all capture processes. Impurities constitute from 15 vol% to 0.05 vol%. Removing the impurities is possible with subsequent gas purification [34]. That said, it should be remembered that the employed CO_2 separation techniques may also supply impurities. In the most commonly used technology, which is the absorption technology, the solvent used for separating CO_2 from the combustion gas is MEA. MEA may undergo degradation, as a result of which the following products can form: ammonia, aldehydes and carboxylic acid degradation products, formate (HCOO–), hydroxyethyl-formamide (HEF) and hydroxyethyl-imidazole (HEI). Another problem associated with MEA degradation is acid gases, such as SO_x, which will react irreversibly with MEA to form corrosive salts. Therefore, it is mandatory to remove SO_x before CO_2 separation. Moreover, sulfur trioxide (SO_3) will yield both thermostable salts, as well as corrosive H_2SO_4 aerosol in the scrubbers. Other impurities that can cause MEA degradation are fly ash, soot and NO_x compounds, which create a problem similar to that caused by SO_x. Fly ash causes a direct degradation of the solvent, though it may also cause difficulties in the SO_2, scrubber, which will have an indirect effect on MEA decomposition [34].

2.5.2 Impurities Present in Gas Streams from Energy-Intensive Industries

Data relating to the composition of CO_2 streams originating from industries other than the power industry are hard to find in most of the industries, therefore they often rely on estimations [8]. This concerns chiefly the compositions CO_2 streams generated as a result of CO_2 capture from major non-energy CO_2 emitters, including oil refineries, cement plants, and coke and lime production plants. Metallurgical processes – impurities in CO_2 streams captured from ironworks and steel mills differ considerably. Last and Schmick [8] have estimated that the composition of CO_2 captured from ironworks and steel mills is similar to that of CO_2 captured from coal-fired power stations. Technologies used for the capture of CO_2 from iron and steel production include pressure-swing adsorption (PSA) and alkanolamine absorption vessels. Impurities coming from metallurgical processes are similar to those from combustion processes. The main impurities generated by those processes are therefore sulfur and nitrogen oxides [3].

2.5.2.1 Cement production

NO_x, SO_2, CO and CO_2 are the main gas emissions from PC production. Slight amounts of VOCs, ammonia (NH_3), chlorine and HCl may also be emitted. Emissions can also comprise incomplete combustion products that are considered hazardous. As some facilities burn waste fuels, they may also emit small amounts of additional hazardous organic waste impurities. Moreover, raw-material and fuel charges normally contain traces of heavy metals that can be emitted in the form of particles or vapors [3].

2.5.2.2 Lime production

Lime is produced by the calcination of limestone, dolomite or other minerals, most commonly in rotary kilns. Combustion gas containing CO_2, CO, SO_2 and NO_x is generated in lime kilns, and their emissions are influenced by the properties of fuels used for heating the kiln, the properties of the charge mineral material, the quality of lime produced, the type of kiln employed and the type of impurity monitoring instrumentation used. Toxic components of lime kiln combustion gas are metals, such as arsenic, cadmium, chromium, nickel, as well as HCl. Lime kiln waste gas contains around 50% CO_2 [3, 8].

2.5.2.3 Natural gas processing

Natural gas is normally subjected to processing prior to being delivered on the market. Natural gas tanks contain significant quantities of impurities, mainly CO_2 and H_2S, therefore additional treatment should be applied to remove those impurities via ammine or membrane separation. As a result of those processes, high-purity CO_2 streams form, which can be stored. In the case of CO_2 in natural gas processing, co-absorption of hydrocarbons and H_2S can be an issue [3]. To sum up, waste CO_2 streams with a varying CO_2 content and varying composition are emitted by different industries, including power stations, cement plants, ironworks and still mills, refineries and chemical plants constitute a potential source of the useful reactant.

2.6 CARBON DIOXIDE PURITY REQUIREMENTS

In order to transport, store or utilize CO_2 captured from the power industry or other energy-intensive industries, limits of impurities are necessary, which will classify a given type of stream to a specific application. Literature presents studies [31–41] which provide recommended limits of impurities for individual components of the CO_2 stream for the subsequent utilization or storage of the CO_2. The presence of impurities in a CO_2 stream may shift the boundaries in the CO_2 phase diagram toward higher pressures, which means that in order to maintain CO_2 in the dense phase, higher operational pressures will be needed.

2.6.1 Carbon Dioxide Purity Requirements – For Applications Involving Pipeline Transport and Storage

It is recommended that the total concentration of non-condensing components of air origin (N_2, O_2 and Ar) does not exceed 4% due to their impact on compression and transport cost. Moreover, these components may reduce the capacity for retaining the structural CO_2 in geological formations. Hydrogen can be present in CO_2 streams coming from the pre-combustion CO_2 capture process. EOR applications require more stringent limits, especially as far as the presence of O_2 is concerned, which should be kept below 100 ppm, as it favors the growth of microorganisms in reaction with hydrocarbons. The restrictions for the presence of water have been established to restrain corrosion caused by the in-situ formation of carbonic acid. Sulfur compounds (H_2S, COS, SO_2 and SO_3) create the risk of corrosion in the presence of water and should therefore be removed to a certain level, and on top of that, there are also some additional concerns about the toxicity of H_2S. NO_x compounds present in CO_2 streams may pose the risk of corrosion due to the formation of nitric acid. For CO_2 streams originating from CCS, the limit 100 ppmv has been proposed. Among the trace metals that may be present in CO_2 streams, mercury is of key importance due to its toxicity and corrosive effect on many metals. In view of its toxicity, limits are suggested also for carbon dioxide, but these are very diverse in the literature. The removal of particulate solids from CO_2 streams results from the need for preventing damage to, or contamination of the equipment. Design parameters for particulates are given as 0–1 ppmv. For other components which can be present in CO_2 streams (e.g., HF, NH_3, MEA, Selexol), too little information is available to understand their subsequent effect on the transport and storage and to determine their maximum permissible quantities [3, 31–41].

2.6.2 Carbon Dioxide Purity Requirements – For Applications Involving Utilization

Requirements for CO_2 stream purity must be adjusted to the needs of a specific utilization method. Some of them are more stringent than others and require more purified waste CO_2 streams. Table 2.2 provides carbon dioxide purity requirements for different branches of industry, e.g. mineral carbonation, chemicals and fuels (conversion), biological carbon dioxide utilization, EOR, greenhouses and the food and beverage industry.

An advantage of mineral carbonation as one of the methods of CO_2 utilization is not very high requirements for waste CO_2 stream purity. Impurities, such as NO_x and SO_x, present in a CO_2 stream originating from industrial applications do not affect the carbonation reaction. Less stringent requirements for CO_2 stream purity in the case of industrial application contribute to a lower consumption of energy, e.g., for CO_2 stream purification. In addition, CO_2 bonded in carbonates can be stored for a long time in a

TABLE 2.2
Carbon dioxide purity requirements for different branches of industry

Branches of industry	Purity requirements	Ref.
EOR	• ($O_2 < 100$ ppm, $NO_x < 100$ ppm, particulates < 0–1 ppm)	[3, 41, 42]
Mineral carbonation/ construction materials	• Not very high requirements for waste CO_2 stream purity. • Impurities, such as NO_x and SO_x, present in a CO_2 stream originating from industrial applications do not affect the carbonation reaction. • Industrial sources of CO_2 streams in high CO_2 concentration will be advantageous, as a higher CO_2 concentration favors the kinetics of the process. • Possibility of using CO_2 streams with low CO_2 concentration <25%.	[3, 39]
Chemicals and fuels conversion	• The catalysts must be selected to suit the impurities contained in the waste CO_2 stream and be able to operate at a high CO_2 concentration in the waste gas stream. • Even a small concentration of, e.g., hydrogen sulfide totally deactivates the catalysts. • Sulfur and arsenic compounds are the most typical poisons for metals in hydrogenation, dehydrogenation and steam reforming reactions. • Arsine was a strong poison for the synthesis of methanol, and with its content at a level of 150 ppbv a rapid deactivation of the catalyst occurred. • More advanced and costly purification methods are required • 3.2 ppm H_2S in synthesis gas reduce the rate of methanol formation by 50%.	[3, 31–41]
Methane and biological conversion	• Biological conversion has a considerable advantage over other technologies, as it does not require purified CO_2 streams. • Algae biomass successfully absorbs CO_2 from coal-fired power station flue gas, in which CO_2 concentration ranges from 12 to 15 mol%. • Flue gas from natural gas-fired power stations can be used for this purpose, where waste streams with low contents of sulfur and nitrogen are generated, because the low NO_x and SO_x levels in those streams can be metabolized by the majority of algae strains. • Flue gas coming from coal-fired power stations presents a challenge, as the CO_2 stream acquired from it contains additional impurities, such as arsenic, cadmium, mercury and selenium. • Coal-fired power station flue gas, containing more impurities, will be more suitable for the production of biofuels than for the production, e.g., protein for animal fodder.	[3]
Greenhouse, food industry, drinks	• Purified CO_2 streams are required due to high purity requirements. • Processes are being searched for, which could valorize the impurities contained in the CO_2 streams.	[3, 39]

safe manner. In the process of mineral carbonation, industrial sources of CO_2 streams in high CO_2 concentration will be advantageous, as a higher CO_2 concentration favors the kinetics of the process. However, because of the fact that in the majority of waste CO_2 streams the CO_2 content is below 25 vol% CO_2, these processes will require either inexpensive CO_2 capture systems or a capability to handle large quantities of gas streams containing components other than CO_2. An issue in the case of mineral carbonation is also the fact that diluted CO_2 is most often available a long way from major building markets, which might require that either mineral carbonation plants be located next to CO_2 emission sites or CO_2 be transported to production sites, which would generate additional costs associated with CO_2 transportation. Therefore, in order to develop mineral carbonation, it is necessary to maximize the utilization of diluted CO_2 streams, as acquiring concentrated and purified CO_2 would heavily affect the costs of building materials production [3]. In the process of chemicals and fuels conversion [3] to transform CO_2 into fuels or chemicals, the main challenges are: to minimize the energy consumption and improve the selectivity and stability of the catalyst, and to assess the tolerance of the catalyst to impurities present in waste CO_2 streams. In view of the fact that most of the systems converting CO_2 into fuels or chemicals tested on a small scale use purified CO_2, there is not sufficient information on the behavior of the catalyst in the presence of a non-purified CO_2 stream that might contribute to its fast decomposition. In the case of using waste CO_2 as a raw material, the catalysts must be selected to suit the impurities contained in the waste CO_2 stream and be able to operate at a high CO_2 concentration in the waste gas stream. A solution is also to pre-concentrate and/or pre-clean the gas, which will obviously affect the economy of the process [3]. There is a need for research on the interface between CO_2 conversion systems and the capture technology, or developing catalysts that will be able to react directly with captured CO_2. Hendrikson et al. [5] have examined the effect of monoethanolamine (MEA), an absorbent commonly used for CO_2 separation, on catalysts. Even a small concentration of, e.g., hydrogen sulfide totally deactivates the catalysts. The effect of impurities on heterogeneous catalysts was studied by Bartholomew [43] who has classified sulfur and arsenic compounds as the most typical poisons for metals in hydrogenation, dehydrogenation and steam reforming reactions. He found that arsine was a strong poison for the synthesis of methanol, and with its content at a level of 150 ppbv, a rapid deactivation of the $CuO/ZnO/Al_2O_3$ catalyst occurred. The deactivation was caused by the dissociative adsorption of arsenic on the Cu surface with the formation of gaseous H_2 and Cu_3As. Research has also shown that the effect of nitrogen compounds on metallic catalysts is less severe than the effect of sulfur compounds. This problem was also noticed by [7], who stated that if CO_2 to be used for methanol synthesis is taken from industrial combustion gas derived from the power industry or other energy-intensive industries, it is an important consideration to examine the tolerance of the employed methanol synthesis catalyst, In_2O_3/ZrO_2, to typical impurities present in industrial CO_2 sources. In conjunction with the fact that CO_2 deriving from industrial sources contains, depending on its origin, contaminants, such as sulfur and nitrogen compounds (H_2S, SO_x or NO_x), hydrogen halides and hydrocarbons, it must be previously purified using absorption technologies to be able to be used in catalytic hydrogenation processes. Sometimes, more advanced and costly purification methods are required, if the CO_2 hydrogenation catalyst to be used in subsequent synthesis is prone to poisoning by sulfur or nitrogen compounds. In the process of biological carbon dioxide utilization, algae biomass successfully absorbs CO_2 from coal-fired power station flue gas, in which CO_2 concentration ranges from 12 to 15 mol%. This is confirmed by several operating pilot plants, where on-site generated flue gas is simultaneously used for supplying CO_2 for growing algae. Such location strategies are extremely advantageous, as they eliminate the need for transporting CO_2 through pipelines, which would require the purification and concentration of the CO_2 captured from the source. This technology makes also use of the natural flexibility of microorganisms' photosynthesis in using streams in varying CO_2 concentrations. The utilization of biomass using power plant CO_2 provides opportunities for capturing carbon dioxide on a large scale. Also, flue gas from natural gas-fired power stations can be used for this purpose, where waste streams with low contents of sulfur and nitrogen are generated, because the low NO_x and SO_x levels in those streams can be metabolized by the majority of algae strains. In turn, flue gas coming from coal-fired power stations presents a challenge, as the CO_2 stream acquired from it contains additional impurities, such as arsenic, cadmium, mercury and selenium [3]. The contents of those impurities may limit the potential uses of biomass products. Therefore, coal-fired power station flue gas, containing more impurities, will be more suitable for the production of biofuels than for the production, e.g., protein for animal fodder. According to [3], biological conversion has a considerable advantage over other technologies, as it does not require purified CO_2 streams. However, co-location with CO_2 sources would be advisable in this technology due to transportation barriers. The production of photosynthetic alga biomass has also numerous advantages, compared to conventional types of growing, in terms of the carbon footprint, water consumption footprint and protein content. The alga protein offers a chance to supplement of, or substitute for conventional crops as a source of animal fodder and/or human food; this requires, however, investigations on a larger scale. Of course, one should be aware of the fact that CO_2 streams acquired from flue gas, containing heavy metal pollutants, would be unsuitable to be used in animal or human food. In applications, such as greenhouses, the food and beverage industry, and especially food processing [3], purified CO_2 streams are required due to high purity requirements. However, processes are being searched for, which could valorize the impurities contained in the CO_2 streams. An example can be LanzaTech that makes use of microbes to convert carbon-rich waste gas containing carbon monoxide, hydrogen, carbon dioxide, methane and other species into various products. The microbes use carbon monoxide as a source of energy [3]. In the utilization of CO_2, it is crucial to adapt the source of origin of the CO_2 stream to the place and method of its utilization. This requires the integration of CO_2 capture and separation with the method of utilization. Due to the fact of occurring CO_2 streams of a varying CO_2 content and different contents of impurities, they should be matched with utilization methods that require CO_2 streams of either a higher or lower CO_2 content and allow or not the presence of impurities. Some of the utilization methods require pure CO_2, free from impurities that could upset the utilization process and lower the quality of the product. Combustion gas rich in CO_2 may not be used directly in some applications, such as food processing or conversion into chemicals, because of other impurities, such as NO_x and SO_x, contained in them, but may, in turn, be used in other applications, such as EOR, which do not require such

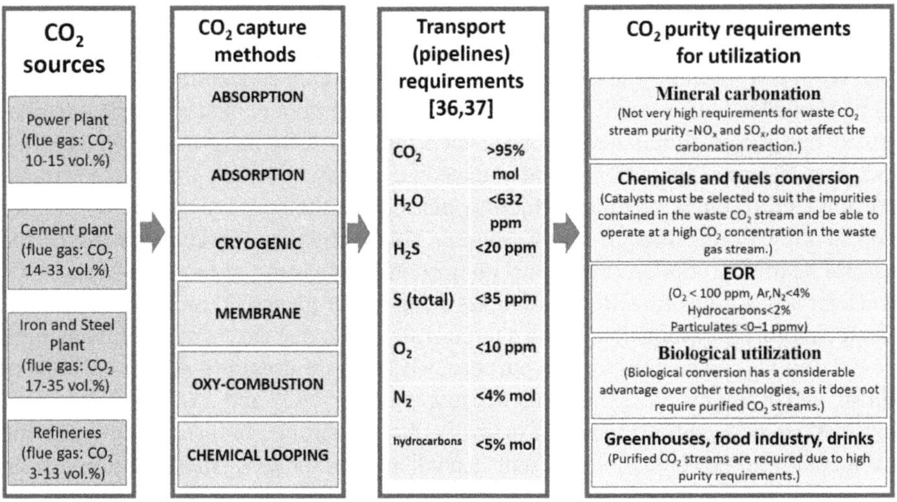

FIGURE 2.6 CO_2 quality requirement for a system with CO_2 utilization.

restrictive compositions. Each additional purification of a waste CO_2 increases the energy consumption. Matching the source of CO_2 with the method of its utilization requires, therefore, analyses to select the appropriate CO_2 utilization method that will minimize the energy consumption and greenhouse gas emissions involved with CO_2 capture and take into account the distance from the source and the transport of CO_2 between the source and the place of utilization [2]. Such analyses are essential for the recommendation of any possibilities of utilizing CO_2 captured from the power industry and other industries (Figure 2.6).

For many utilization technologies that are at an early stage of development, the required purity of carbon dioxide is unknown. With the development of CO_2 utilization technologies, the determination of CO_2 purity levels acceptable by utilization processes will be of key importance [3]. Provided that the requirements for transportation and storage are defined, requirements for the utilization of carbon dioxide are not easy to determine, as there are different applications of carbon dioxide, of which some might not conform to national or international regulations, but do conform to the company's internal practice. Specifications for purity are very well defined in the food industry [10, 39].

REFERENCES

[1] Topham, S.; Bazzanella, A.; Schiebahn, S.; Luhr, S.; Zhao, L.; Otto, A.; Stolten, D. *Ullmann's encyclopedia of industrial chemistry – carbon dioxide chapter*, Wiley-VCH Verlag GmbH & Co. KGaA, **2014**, https://doi.org/10.1002/14356007.a05_165.pub2.

[2] Ghiat, I.; Al-Ansari, T. A review of carbon capture and utilisation as a CO_2 abatement opportunity within the EWF nexus, *Journal of CO_2 Utilization* **2021**, *45*, 101432, https://doi.org/10.1016/j.jcou.2020.101432.

[3] National Academies of Sciences, Engineering, and Medicine **2019**. Gaseous Carbon Waste Streams Utilization: Status and Research Needs. Washington, DC: The National Academies Press, https://doi.org/10.17226/25232.

[4] Schuwer, D.; Arnold, K.; Bienge, K.; Bringezu, S.; Echternacht, L.; Esken, A.; Fischedick, M.; von Geibler, J.; Holler, S.; Merten, F.; Perrey, K.; Pastowski, A.; Pietzner, K.; Schneider, C., Terrapon-Pfaff, J.C.; Vieban, P. *CO_2 ReUse NRW. Evaluating gas sources, demand and utilization for CO_2 and H_2 within the North Rhine-Westphalia area with respect to gas qualities supported by Climate-KIC Deutschland*, Wuppertal Institute for Climate, Environment and Energy, Berlin, **2015**. Accessed September 15, 2021, https://epub.wupperinst.org/frontdoor/deliver/index/docId/6010/file/6010_CO2_ReUse.pdf.

[5] Henriksson, F. *Effect of impurities on hydrogenation of CO_2*, Master's Programme in Chemical, Biochemical and Materials Engineering, **2016**. Accessed September 15, 2021, http://urn.fi/URN:NBN:fi:aalto-201612085813.

[6] Brown, T.; Gambhir, A.; Florin, L.; Fennel, P. *Reducing of CO_2 emissions from heavy industry: a review of technologies and considerations for policy makers*, Imperial College, London, **2012**. Accessed September 15, 2021, https://www.imperial.ac.uk/media/imperial-college/grantham-institute/public/publications/briefing-papers/Reducing-CO2-emissions-from-heavy-industry---Grantham-BP-7.pdf.

[7] Schühle, P.; Schmidt, M.; Schill, L.; Riisager, A.; Wasserscheida, P.; Albert, J. Influence of gas impurities on the hydrogenation of CO_2 to methanol using indium-based catalysts, *Catalysis Science & Technology* **2020**, *10*, 7309, https://doi.org/10.1039/D0CY00946F.

[8] Last, G.V.; Schmick, M.T. *Identification and selection of major carbon dioxide stream compositions*, Prepared for the U.S. Department of Energy under Contract DE-AC05–76RL01830, Pacific Northwest National Laboratory Richland, Washington, 99352, **2011**. Accessed September 15, 2021, https://www.pnnl.gov/main/publications/external/technical_reports/PNNL-20493.pdf.

[9] Fischedick, M.; Roy, J.; Abdel-Aziz, A.; Acquaye, A.; Allwood, J.M.; Ceron, J.P.; Geng, Y.; Kheshgi, H.; Lanza, A.; Perczyk, D.; Price, L.; Santalla, E., Sheinbaum, C.; Tanaka, K. Industry. In *Climate change 2014: mitigation of climate change. Contribution of working group III to the fifth assessment report of the intergovernmental panel on climate change*; Edenhofer, O.; Pichs-Madruga, R.; Sokona, Y.; Farahani, E.; Kadner, S.; Seyboth, K.; Adler, A.; Baum, I.; Brunner, S.; Eickemeier, P.; Kriemann, B.; Savolainen, J.; Schlömer, S.; von Stechow, C.; Zwickel, T.; Minx, J.C. Eds.; Cambridge University Press, Cambridge and New York, **2014**. Accessed September 15, 202, https://www.ipcc.ch/site/assets/uploads/2018/02/ipcc_wg3_ar5_chapter10.pdf.

[10] Murugan, A.; Brown, R.J.C.; Wilmot, R.; Hussain, D.; Bartlett, S.; Brewer, P.J.; Worton, D.R.; Bacquart, T.; Gardiner, T.; Robinson, R.A.; Finlayson, A.J. Performing quality assurance of carbon dioxide for carbon capture and storage, *C Journal of Carbon Research* **2020**, *6*, 76, https://doi.org/10.3390/c6040076.

[11] Bhown, A.S.; Freeman, B.C. Analysis and status of post-combustion carbon dioxide capture technologies, *Environmental Science & Technology* **2011**, *45*, 8624–8632, https://doi.org/10.1021/es104291d.

[12] Gomes, V.G.; Yee Kevin, W.K. Pressure swing adsorption for carbon dioxide sequestration from exhaust gases. *Separation and Purification Technology* **2002**, *28*, 161–171, https://doi.org/10.1016/S1383-5866(02)00064-3.

[13] Suzuki, T.; Sakoda, A.; Suzuki, M.; Izumi, J. Recovery of carbon dioxide from stack gas by piston-driven ultra-rapid PSA. *Journal of Chemical Engineering of Japan* **1997**, *30*, 1026–1033, https://doi.org/10.1252/jcej.30.1026.

[14] Ishibashi, M.; Ota, H.; Akutsu, N.; Umeda, S.; Tajika, M.; Izumi, J.; Yasutake, A.; Kabata, T.; Kageyama, Y. Technology for removing carbon dioxide from power plant flue gas by the physical adsorption method. *Energy Conversion and Management* **1996**, *37*, 929–933, https://doi.org/10.1016/0196-8904(95)00279-0.

[15] Ling, J.; Ntiamoah, A.; Xiao, P.; Xu, D.; Webley, P.A., Zhai, Y. Overview of CO_2 capture from flue gas streams by vacuum pressure swing adsorption technology. *Austin Journal of Chemical Engineering* **2014**, *1*, 1009.

[16] Majchrzak-Kucęba, I. *High-efficiency adsorption technology based on advanced CO_2 sorbents for near zero emission from energy and other industrial plants*, Publishing Office of CUT, Poland, **2016**.

[17] Wawrzyńczak, D.; Majchrzak-Kucęba, I.; Srokosz, K.; Kozak, M.; Nowak, W.; Zdeb, J.; Smółka, W.; Zajchowski, A. The pilot dual-reflux vacuum pressure swing adsorption unit for CO_2 capture from flue gas, *Separation and Purification Technology* **2019**, *209*, 560–570, https://doi.org/10.1016/j.seppur.2018.07.079.

[18] Majchrzak-Kucęba, I.; Wawrzyńczak, D.; Ściubidło, A.; Zdeb, J.; Smółka, W.; Zajchowski, A. Stability and regenerability of activated carbon used for CO_2 removal in pilot DR-VPSA unit in real power plant conditions, *Journal of CO_2 Utilization* **2019**, *29*, 1–11, https://doi.org/10.1016/j.jcou.2018.11.003.

[19] Majchrzak-Kucęba, I.; Wawrzyńczak, D.; Zdeb, J.; Smółka, W.; Zajchowski, A. Treatment of flue gas in a CO_2 capture pilot plant for a commercial CFB boiler, *Energies* **2021**, *14*, 2458, https://doi.org/10.3390/en14092458.

[20] *Putting CO_2 to use, creating value from emissions*, IEA Publications, **2019**. Accessed September 15, 2021, https://www.iea.org/reports/putting-co2-to-use.

[21] Anwar, M.N.; Fayyaz, A.; Sohail, N.F.; Khokhar, M.F.; Baqar, M.; Yasar, A.; Rasool, K.; Nazir, A.; Raja, M.U.F.; Rehan, M.; Aghbashlo, M.; Tabatabaei, M.; Nizami, A.S. CO_2 utilization: turning greenhouse gas into fuels and valuable products, *Journal of Environmental Management* **2020**, *260*, 110059, https://doi.org/10.1016/j.jenvman.2019.110059.

[22] Billiga, N.E.; Deckerb, M.; Benzingerc, W.; Ketelsena, F.; Pfeiferc, P.; Petersb, R.; Stoltena, D.; Thräna, D. Non-fossil CO_2 recycling – the technical potential for the present and future utilization for fuels in Germany, *Journal of CO_2 Utilization* **2019**, *30*, 130–141, https://doi.org/10.1016/j.jcou.2019.01.012.

[23] Pérez-Fortes, M.; Schöneberger, J.C.; Boulamanti, A.; Tzimas, E. Methanol synthesis using captured CO_2 as raw material: techno-economic and environmental assessment, *Applied Energy* **2016**, *161*, 718–732, https://doi.org/10.1016/j.apenergy.2015.07.067.

[24] Liu, B.; Qin, J.; Shi, J.; Jiang, J.; Wua, X.; He, Z. New perspectives on utilization of CO_2 sequestration technologies in cement-based materials, *Construction and Building Materials* **2021**, *272*, 121660, https://doi.org/10.1016/j.conbuildmat.2020.121660.

[25] Meunier, N.; Chauvy, R.; Mouhoubi, S.; Thomas, D.; Weireld, G.D. Alternative production of methanol from industrial CO_2, *Renewable Energy* **2020**, *146*, 1192–1203, https://doi.org/10.1016/j.renene.2019.07.010.

[26] Duttan, A.; Farooq, S.; Karimi, I.A.; Khan, S.A. Assessing the potential of CO_2 utilization with an integrated framework for producing power and chemicals, *Journal of CO_2 Utilization* **2017**, *19*, 49–57, https://doi.org/10.1016/j.jcou.2017.03.005.

[27] Zhang, Y.; Wang, R.; Liu, Z.; Zhang, Z. A novel carbonate binder from waste hydrated cement paste for utilization of CO_2, *Journal of CO_2 Utilization* **2019**, *32*, 276–280, https://doi.org/10.1016/j.jcou.2019.05.001.

[28] Stuardi, F.M.; MacPherson, F.; Leclaire, J. Integrated CO_2 capture and utilization: a priority research direction, *Current Opinion in Green and Sustainable Chemistry* **2019**, *16*, 71–76, https://doi.org/10.1016/j.cogsc.2019.02.003.

[29] Bonura, G.; Todaro, S.; Frusteri, L.; Majchrzak-Kucęba, I.; Wawrzyńczak, D.; Paszti, Z.; Talas, E.; Tompos, A.; Ferenc, L.; Solt, H.; Cannilla, C.; Frusteri, F. Inside the reaction mechanism of direct CO_2 conversion to DME over zeolite-based hybrid catalysts, *Applied Catalysis B: Environmental* **2021**, *294*, 120255, https://doi.org/10.1016/j.apcatb.2021.120255.

[30] Saravanan, A.; Kumar, P.S.; Dai-Viet, N.V.; Jeevanantham, S.; Bhuvaneswari, V.; Anantha N.V.; Yaashikaa, P.R.; Swetha, S.; Reshma, B. A comprehensive review on different approaches for CO_2 utilization and conversion pathways, *Chemical Engineering Science* **2021**, *236*, 116515, https://doi.org/10.1016/j.ces.2021.116515.

[31] Porter, R.T.J.; Fairweather, M.; Pourkashanian, M.; Woolley, R.M. The range and level of impurities in CO_2 streams from different carbon capture sources, *International Journal of Greenhouse Gas Control* **2015**, *36*, 161–174, https://doi.org/10.1016/j.ijggc.2015.02.016.
[32] Lee, J.Y.; Keener, T.C.; Yang, J. Potential flue gas impurities in carbon dioxide streams separated from coal-fired power plants, *Journal of Air & Waste Management Association* **2009**, *59*, 725–732, https://doi.org/10.3155/1047–3289.59.6.725.
[33] Zakkour, P.; Cook G. *Roadmap – high purity CO_2 sources: final draft sectoral assessment*, Carbon Counts Company (UK) Ltd, **2010**, https://doi.org/10.13140/RG.2.1.3717.8722.
[34] Al-Siyabi, I. Effect of impurities on CO_2 stream properties, Heriot-Watt University Institute of Petroleum Engineering February, **2013**, Doctoral Thesis (Energy, Geoscience, Infrastructure and Society). Accessed September 15, 2021, http://hdl.handle.net/10399/2643.
[35] Anheden, M.; Andersson, A.; Bernstone, C.; Eriksson, S.; Yan, J.; Liljemark, S.; Wall, C. CO_2 quality requirement for a system with CO_2 capture, transport and storage. In *Greenhouse gas control technologies 7*; Wilson, E.S.; Rubin, D.W.; Keith, C.F.; Gilboy, M.; Thambimuthu, T.; Morris, J.; Gale, K. Eds.; Elsevier Science Ltd, 2559–2564, https://doi.org/10.1016/B978-008044704-9/50373-6.
[36] Pipitone, G.; Bolland, O. Power generation with CO_2 capture: technology for CO_2 purification, *International Journal of Greenhouse Gas Control* **2009**, *3*, 528–534, https://doi.org/10.1016/j.ijggc.2009.03.001.
[37] Walspurger, S.; Van Dijk, H.A.J. *Edgar CO_2 purity. Type and quantities of impurities related to CO_2 point source and capture technology. A literature study*, **2012**, ECN-E-12–054. Accessed September 15, 2021, https://publicaties.ecn.nl/PdfFetch.aspx?nr=ECN-E--12-054.
[38] Shirley, P., Myles, P. *Quality guidelines for Energy system studies. CO_2 impurity designed parameters.* NETL National Energy Technology Laboratory, NETL-PUB-22529, **2019**. Accessed September 15, 2021, https://www.netl.doe.gov/energy-analysis/details?id=3740.
[39] *Carbon dioxide food and beverages grade, source qualification, quality standards and verification*, EIGA Doc 70/17. Revision of Doc 70/08. Prepared by EIGA WG-8 Food Gases and Carbon Dioxide.
[40] Novel carbon capture and utilisation technologies research and climate aspects. SAPEA Evidence Review Report No. 2. Informs the European Commission Group of Chief Scientific Advisors Scientific Opinion No. 4/2018. Accessed September 15, 2021, https://www.sapea.info/wp-content/uploads/CCU-report-web-version.pdf.
[41] Shah, M.M. Carbon dioxide (CO_2) compression and purification technology for oxy-fuel combustion. In *Oxy-fuel combustion for power generation and carbon dioxide (CO_2) capture*; Zheng, L. Eds., Woodhead Publishing Series in Energy, Elsevier: 17 (11), **2011**, 228–256.
[42] Shah, M.; Degenstein, N.; Zanfir, M.; Solunke, R.; Kumart, R.; Bugayong, J.; Burgers, K. Purification of oxy-combustion flue gas for SO_x/NO_x removal and high CO_2 recovery, Praxair, Inc., presentation at 2nd Oxyfuel Combustion Conference, Yeppoon, Australia, September 12–16, **2011**. Accessed September 15, 2021, https://ieaghg.org/docs/General_Docs/OCC2/Abstracts/Abstract/occ2Final00151.pdf.
[43] Bartholomew, C.H. Mechanisms of catalyst deactivation. *Applied Catalysis A: General* **2001**, *212*, 17–60, https://doi.org/10.1016/S0926-860X(00)00843-7.

3 Adsorption technology for CO_2 capture

Dariusz Wawrzyńczak

CONTENTS

3.1 INTRODUCTION

Adsorption is a well-known method of removing constituents or pollutants from a gaseous or liquid environment. Although the adsorption phenomenon was noticed already in ancient times (and applied for reduction of copper, zinc and tin ores, medicinal purposes, purification of drinking water using charcoal), only the end of the eighteenth century has brought the first quantitative observations (gas uptake, decolorization of tartaric acid solutions, removal of colors from sugar using charcoal) [1]. The following years have brought the development of adsorption science by ongoing interplay between theory and experiment, but pioneering development has started in the twentieth century: commercial development of activated carbon processing and its application diversities, development of various theories as well as invention of zeolite synthesis technology [1].

Currently, adsorption is used in many industrial applications, for instance to: gas separation, as well as recovery processes-production of oxygen and nitrogen (from air), separation of paraffins, drying or dehydration of different streams (e.g. air, natural gas, olefin, solvents), purification of hydrogen (from steam reformer products), purification of natural gas (from carbon dioxide) as well as landfill gas, aromatic separation, water purification, desulphurization, etc. [2, 3]. Last time, due to environmental issues related to counteracting greenhouse gas emissions, the adsorption technology is also proposed and investigated for carbon dioxide capture from flue gases, next to absorption, membranes and cryogenics.

3.2 FUNDAMENTALS OF ADSORPTION TECHNOLOGY

Adsorption is the retention of a substance (being in the gas or liquid phase) onto the surface or inside the volume of micropores of a solid (called adsorbent), as a result

DOI: 10.1201/9781003336587-3

of the intermolecular forces. This attraction of the molecules can be realized from a gaseous or liquid environment. The substance which is being adsorbed is called adsorptive and after retention – adsorbate. The useful product or products can be obtained either in adsorption step (as a raffinate product which is weakly removed from substance by the adsorbent) or in regeneration step (as an extract product which desorbates from the adsorbent) or from both steps.

Adsorption separation can be achieved by three distinct mechanisms [2, 4, 5]:

- Steric – in which pores dimension of adsorbent limits specific size of molecules to enter inside the porous material, characteristic for molecular sieving property of zeolites.
- Kinetic – where mass transfer resistance is crucial (differences in diffusion rates of different molecules), characteristic for molecular sieve carbon.
- Equilibrium – which depends on affinities of the material to attract the molecules (the stronger adsorbing species are preferentially removed by the solid).

Due to the nature of the bonding forces, the adsorption process can be divided into: a physical adsorption (physisorption), characterizing van der Waals interactions and electrostatic forces, between the adsorbate and the surface of the adsorbent, and a chemical adsorption (chemisorption), involves formation of chemical bond between the adsorbate and the adsorbent [2, 3, 5]. The first process, due to low heat of adsorption (generally below 63 kJ/mol), makes it easily reversible, the second one needs activation, may be slow and irreversible [2, 3]. In turn, depending on the concentration of the component being strongly adsorbed from the mixture, the separation process may be divided into purification and bulk separation (the latter one is applied if 10 wt% or more of the mixture is adsorbed) [2].

The adsorption can be realized in different types of reactors [6]:

- Fixed bed where particles of adsorbent are stationary. This type divides into: conventional (beds are filled with loose sorbent which can be different shape and size) and structured beds (adsorbent material is coated with a thin film on the reactor wall – as monolith). The second type maximizes the surface area per volume of adsorbent providing low-pressure drop which improves throughput as compared to fixed bed configurations.
- Moving bed, where sorbent is displaced in the bed. This type divides into: conventional (where reactor is divided into several sections through which the sorbent moves, e.g. an adsorber, a transition, a desorber, a dehydrator, a cooler and a lift) and rotating beds (the bed is in the form of disc divided into sections: adsorption, desorption and cooling).
- Fluidized bed, in which sorbent behaves like a fluid and circulates among adsorber and regenerator. The reactor can be built as one or multistage fluidized bed. The second solution reduces the overall internal back mixing.

Fluidized bed as well as conventional moving bed require mechanically strong particles to minimize attrition, the rotating bed – a good seal between sections of adsorber due to limit the gas leakage while solid is rotating. Therefore, the most systems in use today are fixed-bed, developed as first [7].

3.3 TYPES OF ADSORBENTS

Adsorbents are materials of porous structure and well-developed surface area. The effective adsorbent should be characterized by high selectivity and high capacity [5]. The selectivity depends on equilibrium capacity (which is influenced by specific surface area, contributing to the amount of components being removed from the fluid) as well as kinetic properties (which determines the rate of transfer of the adsorptive to the adsorption sites). Moreover, the adsorbent should have a good mechanical property (strength and resistance to attrition) [8].

The equilibrium effect relies on adsorbate-adsorbent interactions as well as process conditions (temperature, pressure and concentration of gas components) [8] and is characteristic for polar adsorbents which are hydrophilic (corresponds to affinity with polar substances such as water), e.g., aluminosilicates as well as non-polar, being generally hydrophobic (having greater affinity to oil than water), e.g., carbonaceous, polymer or silicate [5, 9].

In turn, the kinetic effect is largely determined by diffusivity of components being separated in micropores [8] and is characteristic for crystalline adsorbents (which micropore dimensions are determined by crystal structure, e.g., zeolites) and amorphous one (which pore size can be adjusted in activation step or by controlled deposition of easy crackable or polymerizable hydrocarbons, e.g., carbon molecular sieves) [5].

Generally, the most known physical adsorbents which can be used in carbon dioxide capture are:

- Activated carbons, well known, produced from variety of sources which contain carbon, and their properties depend on the process production parameters. Activated carbons differ in pore distribution, pore structure, active surface structure and equilibrium capacity. They are characterized by weak physical interactions in relation to the adsorbate (and thus low heat of adsorption) which makes adsorbent regeneration easy [2, 10], good stability and retaining high performance even in the presence of water [11] (see Chapter 4 for more information).
- Zeolites, crystalline nanoporous aluminosilicates of alkali or alkali earth elements such as sodium, potassium and calcium [2]. They are characterized by well-defined size-selective molecular sieve properties [5], high chemical and thermal stability, good CO_2 capacity and selectivity [11]. The presence of water vapor strongly reduces the equilibrium capacity of CO_2 [10] and involves high energy penalty during regeneration [11].
- Metal-organic frameworks (MOFs), three-dimensional organic-inorganic hybrid networks formed by multiple metal-ligand bonds [10] with a porous structure analogous to zeolites [12]. They are characterized by: high surface area, high stability, open channels, permanent porosity [13] as well as show significant advantages in gas selectivity and separation in comparison with traditional adsorbents due to possibility of fine tuning in terms of porosity

and chemical functionality [12]. They are synthesized from the following materials: linkers or bridging ligands (ions or neutral molecules), metal precursors (transition metals and rare earth metals) and the solvents to dissolve the metal salts and ligands which are in solid state [13]. Unfortunately, carbon dioxide capture performance of most MOFs is severely deteriorated by the humidity, due to preferentially and much more strongly bond with H_2O than CO_2 (some of them have a problem in repeated uses under humid conditions) [11].

- Zeolitic imidazolate frameworks (ZIFs), a sub-family of metal-organic structures, combining properties of zeolites and MOFs [14, 15], which consist of inorganic metal ions or metal clusters bridged by imidazolate [16]. They are characterized by highly flexible structure (including pore size and surface properties that can be rationally designed) [16], unexpected thermal (>550°C) and chemical stability [14] that exceeds those of many other MOFs [17]. The humidity, even in trace amounts, has an effect on carbon dioxide capture (but for some of adsorbents, e.g., ZIF-68, the presence of water vapor less than 5% has a negligible effect) [18].
- Covalent organic frameworks (COFs), porous crystalline organic polymers, synthesized by the covalent linkage of organic molecules bonded in a repeating fashion [19], possess almost all the merits of MOFs [11]. They are characterized by: easy pore surface engineering and ordered pore distribution, good stability, high chemical and thermal stability, high CO_2 uptake, great CO_2 selectivity as well as low energy penalty for regeneration [11]. Some of them maintain good CO_2 capture performance even under moist conditions [11].

The production cost of activated carbon as well as zeolites is relatively low, therefore they are the most popular adsorbents in industry. Synthesis of MOFs in bulk scale is limited, due to the challenges in scaling-up of the production as well as high costs of production and environmental, compared to traditional sorbents [12]. In turn, the commercial manufacturing of COFs seems at the moment a distant goal, due to the chemical instability, non-scalability and non-processability [20]. The examples of carbon dioxide equilibrium capacity of some selected sorbents are presented in Table 3.1.

3.4 ADSORBENT REGENERATION METHODS

Each cyclic separation process involves a minimum of two steps: adsorption, in which the preferentially adsorbed species (adsorptives) are picked up from the feed, as well as desorption (called regeneration) during which adsorbates are released from the adsorbent and the regenerated adsorbent is used in the next cycle [5].

In the case of fixed bed, the high-pressure product (purified from the preferentially adsorbed species during adsorption step) is received from the top of the adsorption bed as effluent and called "raffinate" product, while the low-pressure product (received from the bottom of the bed in regeneration step) contains molecules released from the adsorbent in concentrated form and called "extract" product [5].

TABLE 3.1
Examples of CO_2 capture capacities of physical adsorbents

Type of adsorbent	Specific information	CO_2 capacity mmol/g	Temperature °C	Pressure kPa	Ref.
Activated carbon	Activated carbon	2.1	25	100	[10]
	Norit R2030CO2	2.5	25	100	[21]
	Organosorb	2.4	30	100	[22]
Zeolite	13X	4.5	22	100	[10]
	5A	3.4	30	100	[23]
	13X	4.4	30	100	[22]
MOFs	$Cu_3(BTC)_2$	4.1	25	100	[15]
	Mg-MOF-74	8.0	23	100	[20]
	MIL-53(Al)	2.0	30	100	[22]
ZIFs	ZIF-7	2.3	25	100	[15]
	JUC-160	3.5	25	100	[14]
	ZIF-69	3.0	0	100	[19]
COFs	COFJLU-2	4.9	0	100	[19]
	FCTF-1-600	5.5	0	100	[19]
	FCTF-1	4.7	0	100	[19]

In practice, to ensure process continuity, increase in product purity and reduce energy consumption, the following steps of the process (depending on configuration and separation technique used) can be implemented [2, 5]:

- Adsorption/Feed (A) – continuous flow of the feed gas (usually compressed) through the bed during which one or more desired components are removed.
- Pressure equalization (EQ) – single or multi-stage balancing of the pressure between two or more interconnected beds: one after adsorption process is finished and the other after regeneration process is completed (this step allows increase the recovery and saves the energy, the accumulated in the gas of the high-pressure bed).
- Cocurrent depressurization (CD) – single or multi-stage reduction of gas pressure in the bed in the according direction to the feed gas flow (the strong adsorbed species remain in the bed while weak ones are eluted from the saturated zone), used when low-pressure product (extract product) is required at high purity.
- Countercurrent depressurization (CC) – single or multi-stage reduction of gas pressure in the bed in the opposite direction to the feed gas flow, used when high-pressure product (raffinate product) is required in high purity.
- Purge (P) – purge of the bed, following the countercurrent depressurization step, with a portion of high-pressure product (raffinate product) in the

opposite direction than flow of feed gas, to enhance the regeneration of the bed.
- Rinse (R) – purge of the bed, following the adsorption step, with the preferentially adsorbed species at feed pressure in the direction according to the feed flow, to improve low-pressure product (extract product) purity.
- Desorption/Regeneration (D) – conducted by vacuum evacuation (V) or bed temperature increase (PH) to remove strongly adsorbed gas species, ensures high purity of both extract and raffinate products.
- Idle (I) – inactivity step of the bed (usually short), during which no process is carried out.
- Pressurization/Repressurization (R) – increase of the pressure in the bed to the level at the adsorption step with feed gas or raffinate product (improves the purity and recovery of raffinate product).

The adsorption step usually takes place at hypertension (pressure higher than the atmospheric one) and the temperature close to ambient, while the regeneration step can be realized by the following methods [2, 3]:

- Pressure swing – reduction of the pressure in bed to the atmospheric one or even application of vacuum, at essentially constant temperature, usually combined with purging (with raffinate product) or rinsing (with extract product) of the bed (depending on expected product manufacturing) during the regeneration step.
- Thermal (or temperature) swing – heating the bed, usually with a stream of hot gas, e.g., water vapor, feed mixture, raffinate product, to a desorption temperature at which the adsorbate is removed from the bed in the fluid stream (good for strongly adsorbed species).
- Inert purge stripping – relies on the removal of the adsorbate from the bed, without changing the temperature and pressure, and filled the void in the bed with non-adsorbing inert gas, applied when adsorbed species are weakly held, and desorbate is not to be recovered because is significantly diluted by purge gas.
- Displacement gas purge – relies on displacement of the adsorbate (from the adsorbent) by the competitive species, usually at constant temperature and pressure and adsorption in its place (good for strongly held species).

Among the above-regenerated methods, the change of pressure and temperature has found common application in industry. Therefore, the following adsorption techniques are identified:

- PSA – pressure swing adsorption;
- TSA – temperature swing adsorption;
- PTSA – pressure and temperature swing adsorption;
- VSA – vacuum swing adsorption;
- VPSA – vacuum-pressure swing adsorption.

Taking into account the cycle time, the PSA can be realized as:

- RPSA – rapid pressure swing adsorption (cycle time typically 3–10 s, pressure drop in the range of 101–303 kPa) [5] or;
- URPSA – ultra-rapid pressure swing adsorption (cycle time about 0.5 s) [24].

The last of the two, due to high energy requirements (RPSA) and low potential of enriched (URPSA), have not found application in the industry.

Improvement in process efficiency can be achieved by using multiple adsorbent beds with one or more sequences of pressure equalization steps [5], cocurrent/countercurrent depressurization or purge/rinse steps of each adsorber.

The improvements in CO_2 product purity can also be reached by using of two-stage separation process (two installations working in series – first of which enriching the flue gas with CO_2 and the second one is used for its purification) [25, 26]. A similar technique in which there are two stages of separation process is called dual-reflux vacuum pressure swing adsorption (DR-VPSA) [27]. In this case, the adsorber is divided into two sections – first one (upper) that allows to enrich the flue gas stream in CO_2 which next feeds the second (bottom) section where purification process takes place. This partitioning of the adsorber ensures that the part of carbon dioxide from CO_2 enriched gas stream which is not adsorbed in the bottom section, flows through the by-pass to the upper section where its adsorption is continued.

3.5 ADSORPTION TECHNOLOGY IN CO_2 CAPTURE RESEARCH

The adsorption technology is suitable for the feed gas stream which carbon dioxide concentration is starting from 30 vol% [28]. Despite it is tested by many researchers for flue gas which CO_2 concentration starts from 10 vol%, due to: no emission of toxic substances to the atmosphere (solid sorbents are applied), no need to replace sorbent and its necessary to refill [29], high process flexibility, fully automated operation [30] and no aggressive solvents that would cause corrosion of system components.

Apart from many advantages of adsorption technology, there are also some disadvantages, such as the need to ensure a high tightness of the system (especially when vacuum regeneration techniques as VSA or VPSA are applied), treatment of the feed stream (dedust and remove of acid components), gas dehumidification (in the case of PSA/VSA/VPSA techniques the moisture negatively effects on separation effectiveness, especially in the case of hydrophilic sorbents), frequent operation of solenoid valves, the dependence of CO_2 concentration in the feed gas on the purity of the product as well as the recovery rate. Therefore, some process improvements (additional steps of the process, modifications of pressure, temperature, time and feed gas stream, recirculation of low-grade product or two-stage separation) as well as innovations of installation components are introduced. It should be emphasized that the key efficiency factor is also the type of sorbent used. However, to be applied on an industrial scale, it must be economically viable, environmentally neutral, easy produced in large quantities as well as chemically and physically resistant.

Up to date, a lot of laboratory, bench and pilot scale research studies have been carried out on the subject of CO_2 separation from flue gases. Selected experiment results, coming from adsorption installations using different techniques, are summarized in Table 3.2.

The preferred technique of CO_2 separation from flue gas is VSA/VPSA, due to the low demand for power consumption of the blower to slight feed gas compression [32] (as opposed to PSA technique where the post-combustion flue gas must be compressed to high pressure due to low carbon dioxide content, and the compression of flue gas above 150 kPa is not cost-effective [50]). The VSA/VPSA techniques have also advantage over TSA/PTSA ones in the rapid cyclic which leads to much smaller bed sizes [50] as well as no thermal energy required in the sorbent regeneration (preheating) step.

The value of desorption pressure has the substantial effect on two indicators as: product (carbon dioxide) purity as well as recovery rate. The authors [43] proved that the optimum vacuum level depends on the isotherm shape of the adsorbent (e.g. for zeolite 13X is about 4 kPa). However, the deeper vacuum (<5 kPa) requires multistage pump units and large suction line sizes to minimize pressure drop [50] as well as special sealed and large operating valves for low pressure which increases both capital and operating costs [34]. Therefore, due to possible limitations of scalability for large vacuum systems, there are introduced additional separation steps, as for example pressure equalization, which reduces the mechanical work performed by the vacuum pump (to maintain the column pressure at the intermediate level) [34]. Other parameters influenced on product purity are: time of adsorption step (and the associated amount of feed gas introduced into the bed during adsorption step at fixed flow rate affecting the load of the bed) [34], carbon dioxide concentration in the feed gas (which may be increased by recycling the product obtained from the countercurrent depressurization step as well as from fractionation of the extract product or even introducing an additional stage of separation process for pre-enriched product coming from the first section) [38, 40], amount of pressure equalization steps (which reduces the amount of N_2 gas in bed void space before desorption step) [35] and rinsing of the bed (using extract product) [34].

In turn, the factors affecting on recovery rate depend on adsorbent working capacity (kind of adsorbent and applied pressures) [34], time of adsorption step (and associated breakthrough of the bed) [38] and purge/rinse step.

The other indicators, like the productivity and the specific power consumption, are directly influenced by adsorbent properties, process parameters as well as time of the process cycle.

A summary of the effectiveness of adsorption separation process, based on three main indicators: CO_2 purity, recovery and productivity (using the literature analysis in Table 3.2) is presented in Table 3.3.

TABLE 3.2
Comparison of the parameters and results of the adsorption technology of CO_2 capture from the flue gas based on experimental investigations

Scale	Technique	Kind of sorbent	Capacity	Feed gas composition	No of stages/no of beds/no of steps	Process temperature	Process configuration	Characteristics of process parameters	CO_2 purity	CO_2 recovery	Productivity	Specific power consumption	Ref.
			Nm³/h[a]	vol%		°C			%	%	$\frac{kg_{CO_2}}{kg_A h}$	$\frac{kWh}{Mg_{CO_2}}$	
Laboratory	VPSA	Activated carbon	0.57[b]	17% CO_2 79% N_2 4% O_2	I stage: 3-bed 4-step	25–32	(I) A (II) CD (III) D (V) (IV) R	A (152 kPa, 40 s) D (10 kPa, 209 s)	48.8	72.5	0.085[c]	n/a	[31]
					I stage: 3-bed 5-step		(I) A (II) CD (III) RI (IV) D (V) (V) R		70.2	53.1	0.062[c]		
					I stage: 3-bed 6-step		(I) A (II) CD (III) EQ↓ (IV) D (V) (V) EQ↑ (VI) R	A (152 kPa, 22 s) D (10 kPa, 209 s)	68.8	73.9	0.097[c]		
					I stage: 3-bed 8-step		(I) A (II) CD (III) EQ↓ (IV, V) RI (VI) D (V) (VII) EQ↑ (VIII) R		91.2	46.1	0.054[c]		

(Continued)

TABLE 3.2 *(Continued)*

Comparison of the parameters and results of the adsorption technology of CO_2 capture from the flue gas based on experimental investigations

Scale	Technique	Kind of sorbent	Capacity	Feed gas composition	No of stages/no of beds/no of steps	Process temperature	Process configuration	Characteristics of process parameters	CO_2 purity	CO_2 recovery	Productivity	Specific power consumption	Ref.
			Nm^3/h[a]	vol%		°C			%	%	$\frac{kg_{CO_2}}{kg_A h}$	$\frac{kWh}{Mg_{CO_2}}$	
Laboratory	VPSA	Activated carbon	0.11[b]	15% CO_2 85% N_2	I stage: 1-bed 4-step	30	(I) A (II) D (V) (III) PP (IV) R	A (131 kPa, 360 s) D (10 kPa, 135 s) PP (60 s, 0.009 Nm^3/h[d] N_2)	48.56	55.35	0.072[c]	n/a	[32]
								A (203 kPa, 360 s) D (10 kPa, 190 s) PP (60 s, 0.009 Nm^3/h[d] N_2)	57.83	75.48	0.088[c]	n/a	
								A (324 kPa, 360 s) D (10 kPa, 240 s) PP (60 s, 0.009 Nm^3/h[d] N_2)	63.04	96.16	0.106[c]	n/a	
				25% CO_2 75% N_2				A (131 kPa, 360 s) D (10 kPa, 165 s) PP (60 s, 0.009 Nm^3/h[d] N_2)	67.79	55.81	0.116[c]	n/a	

Laboratory	VPSA	Activated carbon	0.06[d]	40% CO_2 60% N_2	I stage (as II stage of separation): 1-bed 4-step	30 (bath thermostated)	(I) A (II) CD (III) D (V) (IV) R	A (203 kPa, 600 s) CD (20 s) D (10 kPa, 430 s) R (180 s)	84.96	86.04	0.072[c]	n/a	[33]
				50% CO_2 50% N_2				A (203 kPa, 600 s) CD (20 s) D (10 kPa, 550 s) R (210 s)	88.75	87.22	0.083[c]	n/a	
				60% CO_2 40% N_2				A (203 kPa, 600 s) CD (20 s) D (10 kPa, 610 s) R (240 s)	94.14	85.08	0.093[c]	n/a	
Laboratory	VSA	Zeolite 13X	0.30[d]	15% CO_2 85% N_2	I stage: 4-bed 20-step	20	(I) A (II) EQ1↓ (III) RI (IV) EQ2↓ (V) EQ3↓ (VI–IX) D (V) (X) EQ3↑ (XI–XIII) I (XIV) EQ2↑ (XV–XVI) I (XVII) EQ1↑ (XVIII) R (RI) (XIX–XX) A	A (110 kPa, 40 s) D (8 kPa, 50 s)	88.04	75.65	0.12	97[c]	[34]
								A (110 kPa, 80 s) D (8 kPa, 90 s)	92.38	64.05	0.08	92[c]	
								A (110 kPa, 120 s) D (8 kPa, 130 s)	95.85	51.27	0.08	86[c]	
								A (110 kPa, 40 s) D (5 kPa, 50 s)	89.20	88.01	0.12	92[c]	
								A (110 kPa, 80 s) D (5 kPa, 90 s)	92.80	80.83	0.13	92[c]	
								A (110 kPa, 120 s) D (5 kPa, 130 s)	96.68	73.77	0.09	89[c]	
Laboratory	VSA	Zeolite 13X	0.30[d]	15% CO_2 85% N_2	I stage: 4-bed 16-step	20	(I) A (II) EQ1↓ (III) EQ2↓ (IV) EQ3↓ (V–VII) D (V) (VIII) EQ3↑ (IX–X) I (XI) EQ2↑ (XII–XIII) I (XIV) EQ1↑ (XV) R (XVI) A	A (108 kPa, 150 s) D (3 kPa, 155 s)	92.4	62.9	0.067	78[c]	[35]
								A (108 kPa, 100 s) D (3 kPa, 105 s)	91.3	77.0	0.088	72[c]	
								A (108 kPa, 40 s) D (3 kPa, 45 s)	79.5	91.3	0.093	64[c]	
								A (108 kPa, 40 s) D (5 kPa, 55 s)	78.8	83.7	0.062	61[c]	

(Continued)

TABLE 3.2 (*Continued*)
Comparison of the parameters and results of the adsorption technology of CO_2 capture from the flue gas based on experimental investigations

Scale	Technique	Kind of sorbent	Capacity	Feed gas composition	No of stages/no of beds/no of steps	Process temperature	Process configuration	Characteristics of process parameters	CO_2 purity	CO_2 recovery	Productivity	Specific power consumption	Ref.
			Nm^3/h[a]	vol%		°C			%	%	$\frac{kg_{CO_2}}{kg_A h}$	$\frac{kWh}{Mg_{CO_2}}$	
Laboratory	VSA	Zeolite 13X	0.20[d]	15% CO_2 85% N_2	I stage: 1-bed 4-step	~30	(I) A (II) D (V) (III) PP (IV) R	A (ambient pressure, 300 s) D (10 kPa, 120 s) PP (10 kPa, 90 s, 0.009 Nm^3/h[d] N_2)	64.1	57.2	0.101	n/a	[36]
								A (ambient pressure, 300 s) D (3 kPa, 300 s) PP (10 kPa, 90 s, 0.009 Nm^3/h[d] N_2)	78.4	78.6	0.133	n/a	
	TSA				I stage: 1-bed 5-step	30 (A) 170 (D) (heatbath)	(I) A (II) D (PH) (III) PP (IV) C + R (V) R	A (~101 kPa, 480 s) D (480 s) PP (90 s, 0.009 Nm^3/h[d] N_2) C (600 s)	91.6	78.1	0.104	n/a	
	VTSA				I stage: 1-bed 6-step	30 (A) 110 (D) (heatbath)	(I) A (II) D (PH) (III) D (V) (IV) PP (V) C + R (VI) R	A (~101 kPa, 600 s) D (PH) (570 s) D (V) (10 kPa, 240 s) PP (10 kPa, 90 s, 0.009 Nm^3/h[d] N_2) C (690 s)	93.6	92.2	0.109	n/a	

								A (~101 kPa, 600 s) D (RH) (570 s) D (V) (3 kPa, 450 s) PP (3 kPa, 90 s, 0.009 Nm^3/h^d N_2) C (690 s)	94.4	98.5	0.120	n/a	
						30 (A) 170 (D) (heatbath)		A (~101 kPa, 660 s) D (RH) (870 s) D (V) (10 kPa, 240 s) PP (10 kPa, 90 s, 0.009 Nm^3/h^d N_2) C (900 s)	94.1	93.8	0.119	n/a	
								A (~101 kPa, 660 s) D (RH) (870 s) D (V) (3 kPa, 300 s) PP (3 kPa, 90 s, 0.009 Nm^3/h^d N_2) C (900 s)	93.5	97.2	0.121	n/a	
Laboratory	VPSA	Zeolite 13X	0.18[d]	15% CO_2 85% N_2	I stage: 1-bed 4-step	90	(I) A (II) CC (III) PP (IV) R	A (130 kPa, 120 s) CC (10 kPa, 70 s) PP (10 kPa, 50 s, 0.03 Nm^3/h^d N_2) R (20 s)	51.54	66.04	0.166[c]	n/a	[37]
								A (130 kPa, 100 s) CC (10 kPa, 70 s) PP (10 kPa, 120 s, 0.03 Nm^3/h^d N_2) R (20 s)	46.17	70.30	0.177[c]	n/a	
			0.12[d]					A (130 kPa, 100 s) CC (10 kPa, 70 s) PP (10 kPa, 70 s, 0.03 Nm^3/h^d N_2) R (30 s)	38.07	86.14	0.145[c]	n/a	

(Continued)

TABLE 3.2 (*Continued*)

Comparison of the parameters and results of the adsorption technology of CO_2 capture from the flue gas based on experimental investigations

Scale	Technique	Kind of sorbent	Capacity	Feed gas composition	No of stages/no of beds/no of steps	Process temperature	Process configuration	Characteristics of process parameters	CO_2 purity	CO_2 recovery	Productivity	Specific power consumption	Ref.
			Nm³/h[a]	vol%		°C			%	%	$\frac{kg_{CO_2}}{kg_A h}$	$\frac{kWh}{Mg_{CO_2}}$	
			0.12[d]		I stage: 1-bed 5-step		(I) A (II) RI (III) CC (IV) PP (V) R	A (130 kPa, 100 s) RI (20 s) CC (10 kPa, 70 s) PP (10 kPa, 70 s, 0.03 Nm³/h[d] N_2) R (30 s)	64.23	84.54	0.142[c]	n/a	
			0.18[d]					A (130 kPa, 100 s) RI (25 s) CC (10 kPa, 70 s) PP (10 kPa, 70 s, 0.03 Nm³/h[d] N_2) R (20 s)	80.90	66.40	0.168[c]	n/a	
Laboratory	VPSA	Activated carbon	0.009	16% CO_2 3.5% O_2 80.5% N_2	I stage: 2-bed 4-step	30	(I) A (II) EQ (III) D (V) (IV) R	A (125 kPa, 900 s) D (3 kPa) R (125 kPa)	59.9	86.9	0.035[c]	n/a	[38]
								A (125 kPa, 900 s) D (10 kPa) R (125 kPa)	55.8	56.6	0.023[c]	n/a	
		Zeolite 13X						A (125 kPa, 900 s) D (3 kPa) R (125 kPa)	78.6	87.6	0.023[c]	n/a	
								A (125 kPa, 900 s) D (10 kPa) R (125 kPa)	72.5	59.2	0.015[c]	n/a	

Laboratory	TSA	Activated carbon	0.002[d]	17% CO_2 83% N_2	I stage: 1-bed 2-step	30 (A) 100 (D) (preheated N_2 gas)	(I) A + C (II) PP + D (PH)	A + C (130 kPa) PP + D (130 kPa, 0.0002 Nm³/h[d] N_2)	43	40	0.035[c]	n/a	[39]
	VSA				I stage: 1-bed 3-step	30	(I) A (II) D (V) (III) R	A (130 kPa) D (0.005 kPa) R (130 kPa)	n/a	87	0.075[c]	n/a	
	VTSA				I stage: 1-bed 3-step	30 (A) 50 (D)	(I) A (II) D (PH) + D (V) (III) R	A (130 kPa) D + H (0.005 kPa) R (130 kPa)	n/a	97	0.084[c]	n/a	
Laboratory	PTSA *(FPTSA)*	Activated carbon	0.0055	15% CO_2 6% O_2 79% N_2	I stage: 2-bed 8-step	~30 (A) ~75 (D) (heating jacket)	(I) A (II) EQ↓ (III) CD (IV) D (PH) (V) PP (VI) C (VII) EQ↑ (VIII) R	A (150 kPa, 2,700 s) PP (100 kPa, 1,350 s, ~0.0001 Nm³/h HPP) C (1,350 s)	39.2 *(RF: 57.4; PF: 29.8)*	42.3 *(RF: 21.2; PF: 21.1)*	0.007 *(RF: 0.004; PF: 0.003)*	n/a	[40]
			0.0057	30% CO_2 10% O_2 60% N_2				A (150 kPa, 2,700 s) PP (100 kPa, 1,350 s, ~0.0001 Nm³/h HPP) C (1,350 s)	68.4 *(RF: 90.6; PF: 61.7)*	42.4 *(RF: 12.7; PF: 29.6)*	0.014 *(RF: 0.004; PF: 0.010)*	n/a	
		Zeolite 13X	0.0054	30% CO_2 10% O_2 60% N_2				A (150 kPa, 2,700 s) PP (100 kPa, 1,350 s, ~0.0001 Nm³/h HPP) C (1,350 s)	86.7 *(RF: 98.6; PF: 77.6)*	65.4 *(RF: 32.1; PF: 33.2)*	0.016 *(RF: 0.008; PF: 0.008)*	n/a	
		Zeolite 13X	0.0054	15% CO_2 6% O_2 79% N_2	I stage: 2-bed 8-step	~50 (A) ~125 (D) (heating jacket)	(I) A (II) EQ↓ (III) CD (IV) D (PH) (V) PP (VI) C (VII) EQ↑ (VIII) R	A (150 kPa, 3,600 s) PP (100 kPa, 1,800 s, ~0.0001 Nm³/h HPP) C (1,800 s)	77.3 *(RF: ≥99.0; PF: 67.6)*	79.9 *(RF: 31.4; PF: 48.5)*	0.010 *(RF: 0.004; PF: 0.006)*	n/a	
			0.0058	30% CO_2 10% O_2 60% N_2				A (150 kPa, 3,600 s) PP (100 kPa, 1,800 s, ~0.0001 Nm³/h HPP) C (1,800 s)	89.3 *(RF: ≥99.0; PF: 75.5)*	78.5 *(RF: 51.7; PF: 26.8)*	0.020 *(RF: 0.013; PF: 0.007)*	n/a	

(Continued)

TABLE 3.2 (*Continued*)

Comparison of the parameters and results of the adsorption technology of CO_2 capture from the flue gas based on experimental investigations

Scale	Technique	Kind of sorbent	Capacity	Feed gas composition	No of stages/no of beds/no of steps	Process temperature	Process configuration	Characteristics of process parameters	CO_2 purity	CO_2 recovery	Productivity	Specific power consumption	Ref.
			Nm^3/h^a	vol%		°C			%	%	$\frac{kg_{CO_2}}{kg_A h}$	$\frac{kWh}{Mg_{CO_2}}$	
Bench	TSA	Zeolite 13X	1.2	10% CO_2 90% N_2	I stage: 1-bed 2-(main) step	Internal heat exchanger for temperature control: - cooling water: 12–15 (A) - steam: 150 (D)	(I) A (II) D (PH)	A (2,940 s) D (960 s)	94	65	0.040[c]	2,444[c]	[41]
		Zeolite 5A					(I) A (II) D (PH)	A (2,940 s) D (960 s)	97	76	0.044[c]	1,972[c]	
							(I) A (II) D (PH)	A (4,440 s) D (1,320 s)	98	83	0.041[c]	1,788[c]	
							(I) A (II) D	A (1,500 s) D (900 s, 0.012 Nm^3/h^c N_2)	94	74	0.051[c]	1,694[c]	
Bench	VSA	Zeolite 13X	7.1[c]	12% CO_2 88% dry air	I stage: 3-bed 6-step	40–50	(I) A (II) EQ1↓ (III) EQ2↓ (IV) D (V) (V) EQ1↑ (VI) EQ2↑	A (130 kPa) E (4 kPa)	82–83	60–80	n/a	96–192[c]	[42, 43]

					I stage: 3-bed 9-step		(I) A (II) EQ1↓ (III) EQ2↓ (IV) EQ3↓ (V) RI (VI) D (V) (VII) EQ3↑ (VIII) EQ2↑ (IX) EQ1↑		90–95	60–70	n/a	144–240[c]	
Pilot	PTSA/PSA	Zeolite Ca-X	1,000	10.8% CO_2 12.9% H_2O 5.1% O_2 71.2% N_2 76 ppm SO_x 60 ppm NO_x 20 mg/Nm^3 dust	I stage: 4-bed PTSA	n/a	n/a	A (105 kPa) D (40 kPa)	40–50	93	n/a		[44, 45]
					II stage: 4-bed PSA	30–50	n/a	A (110–120 kPa) D (10–30 kPa)	99	90		560	
Pilot	PSA	Zeolite 13X	110	10.5% CO_2 (flue gas)	I stage: 2-bed PSA	25–35	n/a	A (110÷120 kPa) D (7 kPa)	40–60	n/a	n/a		[25]
					II stage: 2-bed PSA	15–50	n/a	A (110 ÷ 120 kPa) D (17 kPa)	99	80		642–774	
Pilot	VPSA	Zeolite NaX + NaA (2:1 ratio)	36.7	13% CO_2 79% N_2 8% O_2 (flue gas)	I stage: 4-bed 8-step	22	(I) A (II) RI (III) EQ↓ (IV) D (V) (V) PP (VI) EQ↑ (VII) I (VIII) R	A (120 kPa, 240 s) D (10 kPa, 150 s) PP (75 s)	58.80	91.59	0.107[c]	n/a	[46]

(Continued)

TABLE 3.2 (*Continued*)
Comparison of the parameters and results of the adsorption technology of CO_2 capture from the flue gas based on experimental investigations

Scale	Technique	Kind of sorbent	Capacity	Feed gas composition	No of stages/no of beds/no of steps	Process temperature	Process configuration	Characteristics of process parameters	CO_2 purity	CO_2 recovery	Productivity	Specific power consumption	Ref.
			Nm^3/h[a]	vol%		°C			%	%	$\frac{kg_{CO_2}}{kg_A h}$	$\frac{kWh}{Mg_{CO_2}}$	
Pilot	VPSA	Zeolite	39.8	15.0% CO_2	I stage:	25	(I) A	A (120 kPa, 260 s)	81	86	0.036[c]	733[c]	[47]
		5A		76.5% N_2	3-bed		(II) CD	CD (10 s)					
				8.5% O_2	7-step		(III) RI	RI (240 s,					
				(flue gas)			(IV) EQ↓	1.48 Nm^3/h[d])					
			32.1				(V) D (V)	EQ↓ (10 s)	76	88	0.030[c]	867[c]	
							(VI) PP	D (5 kPa, 190 s)					
							(VII) EQ↑	PP (60 s,					
								1.18 Nm^3/h[d])					
								EQ↑ (10 s)					
Pilot	VPSA	Zeolite	40.3	16.4% CO_2	I stage:	20–40	(I) A	A (117 kPa, 250 s)	82.3	84.7	0.041[c]	564	[48]
		13X			3-bed		(II) CD	CD (10 s)					
					8-step		(III) RI	RI (240 s,					
							(IV) EQ↓	1.18 Nm^3/h[d])					
							(V) D (V)	EQ↓ (10 s)					
							(VI) PP	D (7–8 kPa, 190 s)					
							(VII) EQ↑	PP (60 s,					
							(VIII) R	1.18 Nm^3/h[d])					
								EQ↑ (10 s)					
								R (117 kPa, 10 s)					

Pilot	VPSA/ VPSA	I stage: Zeolite 13X	37.0	16.0% CO_2 (feed gas)	I stage: 3-bed 8-step	30–50	(I) A (II) CD (III) RI (IV) EQ↓ (V) D (V) (VI) PP (VII) EQ↑ (VIII) R	A (118 kPa, 250 s) CD (101.7–101.9 kPa, 10 s) RI (240 s, 1.18 Nm^3/h[d]) EQ↓ (50 kPa, 10 s) D (7–8 kPa, 190 s) PP (60 s, 1.18 Nm^3/h[d]) EQ↑ (10 s) R (118 kPa, 10 s)	79.4	83.7	I stage: 0.037[c]	736*	[26]
		II stage: Activated carbon			II stage: 2-bed 6-step		(I) A (II) CD (III) RI (IV) EQ↓ (V) D (V) (VI) EQ↑	A (125 kPa, 250 s) CD (101.7–101.9 kPa, 10 s) RI (10 s, 1.18 Nm^3/h[d]) EQ↓ (50 kPa, 10 s) D (20 kPa, 270 s) EQ↑ (10 s)	98.8	83.7	II stage: 0.142[c]		
Pilot	VSA	Zeolite 13X	23.57[d]	15% CO_2 85% N_2	I stage: 2-bed 4-step	15–30	(I) A (II) CD (V) (III) D (V) (IV) R	A (150 kPa, 60 s) CD (5.6 kPa, 150 s) D (2.2 kPa, 310 s) R (150 kPa, 40 s)	94.8	89.7	0.073[c]	475	[49]
Pilot	DR-VPSA	I stage: Activated carbon (1 + 2) II stage: Activated carbon (2)	96.1	13.2% CO_2 (flue gas)	I stage: 4-dual bed 8-step	22	(I) A/A (II) EQ↓/EQ↓ (III) I/I (IV) PP/PP (V) D/D (VI) P/P (VII) EQ↑/EQ↑ (VIII) R/R	A (160 kPa, 120 s) EQ (80 kPa) I (80 kPa) PP (45 kPa) E (15 kPa) P (20 kPa) R (160 kPa)	60.5	75.5	0.035[c]	599	[27]

(Continued)

TABLE 3.2 (*Continued*)

Comparison of the parameters and results of the adsorption technology of CO_2 capture from the flue gas based on experimental investigations

Scale	Technique	Kind of sorbent	Capacity	Feed gas composition	No of stages/no of beds/no of steps	Process temperature	Process configuration	Characteristics of process parameters	CO_2 purity	CO_2 recovery	Productivity	Specific power consumption	Ref.
			Nm³/h[a]	vol%		°C			%	%	$\frac{kg_{CO_2}}{kg_A h}$	$\frac{kWh}{Mg_{CO_2}}$	
			97.5	13.8% CO_2 (flue gas)	I/II stage: 4-bed 9-step	18	(I) A1/A2 (II) EQ↓/EQ↓ (III) PP1/E1 (IV) PP2/E2 (V) D/D (VI) P1/P1 (VII) P2/P2 (VIII) EQ↑/EQ↑ (IX) R/R	(I) A1/A2 (160 kPa, 300 s) (II) EQ/EQ (75 kPa) (III) PP1 (65 kPa)/E1 (20 kPa) (IV) PP2 (45 kPa)/E2 (5 kPa) (V) D/D (35 kPa) (VI) P1/P1 (25 kPa) (VII) P2/P2 (15 kPa) (VIII) EQ/EQ (75 kPa) (IX) R/R (160 kPa)	87.5	44.6	0.022[c]	978	

Steps in adsorption technology: A – adsorption; C – cooling; CD – cocurrent depressurization; CC – countercurrent depressurization; D – desorption by: (V) – vacuum; (PH) – preheating; EQ – pressure equalization; I – idle; P – purge; PP – purge and product; R – repressurization; RI – rinse. FPTSA – fractionate pressure and temperature swing adsorption; RF – rich fraction (in CO_2); PF – poor fraction (in CO_2); HPP – high pressure product (raffinate product).

[a] Nm³ (normal cubic meter) – the volume of 1 kmol of gas is 22.4135 m³ (0°C, 101.3 kPa).

[b] Calculated from Sm³ (standard cubic meter) – the volume of 1 kmol of gas is 23.6444 m³ (15°C, 101.325 kPa).

[c] Calculated based on available data in the article.

[d] Calculated from SLPM – the volume of 1 kmol of gas is 22.7105 m³ (0°C, 100 kPa).

TABLE 3.3
The influence of process parameters on effectiveness of adsorption separation technology

Parameter change	CO_2 purity	CO_2 recovery	Productivity
Increase in feed gas stream	↗	↘	↗
Increase in time of adsorption step	↗	↘	↘
Increase in feed gas pressure	↗	↗	↗
Increase in CO_2 concentration in feed gas	↗	↔	↗
Increase in number of separation steps:			
for <9 steps in cycle	↗	↘	↘
for >15 steps in cycle	↗	↔	↗
Increase in time of purge step	↘	↗	↗
Use of 13X instead of AC adsorbent (the same feed gas stream)	↗	↔	↘
Use of 5A instead of 13X zeolite (TSA technique, the same feed gas stream)	↗	↗	↗
Application of two-stage separation process:			
two installations connected in series (in comparison to the same configuration of one stage; adsorbents: (1)13X + (2)AC vs. (1)13X)	↗	↔	↔
DR-VPSA technique (in comparision to the configuration with no bed division, adsorbents: (1)AC + (2)AC)	↗	↘	↘
Technique performance comparison: VSA vs. TSA vs. VTSA	VSA < TSA < VTSA	VSA < TSA < VTSA	TSA < VSA < VTSA

Trend of indicator associated with changing of the process parameter:
↗ – increase.
↘ – decrease.
↔ – comparable level.

3.6 ADSORPTION PROCESS PARAMETERS FOR CASE STUDY

The investigations of carbon dioxide separation from exhaust gases coming from power plant and cement plant were conducted by the Czestochowa University of Technology and the results are summarized in Table 3.4.

The research was conducted in four-bed bench scale installation on simulated exhaust gas, and the experience gained was used during pilot-scale measurement campaigns in real industrial conditions. However, the beginning research has been started from a laboratory scale, and the process of the scaling has been shown in Fig. 3.1.

All installations have been designed and constructed to enable the regulation of feed gas stream, purge gas and gas pressure at individual steps of the process. Applied SCADA provides the opportunity to any configuration (in terms of the number and time of individual steps) of the separation process as well as visualization and archiving of data.

TABLE 3.4
Parameters and results of measurement campaigns of CO_2 separation studies for power and cement plants

Technique used			
Parameter	**Unit**	**DR-VPSA (power plant case study)**	**VPSA (cement plant case study)**
	Inlet		
CO_2	% mol [db]	14.4	29.9
N_2	% mol [db]	62.7[a]	54.7[a]
O_2	% mol [db]	7.7	14.7[a]
Ar	% mol [db]	0.8[a]	0.7[a]
Feed gas flow rate	Nm^3/h	78.1 [27]	2.7
	Outlet		
CO_2	% mol [db]	87.4 [27]	98.1
O_2	% mol [db]	4.7	0.6
N_2	% mol [db]	7.78[a]	1.22[a]
Ar	% mol [db]	0.12[a]	0.08[a]
SO_2	ppm	0–1 [51]	–
NO_x	ppm	27 [51]	–
Recovery	%	50.1 [27]	70.6
Pressure	kPa	100	100
Temperature	°C	20	27
Energy demand	kWh/Mg CO_2	1045[b]	930[b]

[a] Estimated.
[b] Not fully optimized.

Laboratory installations using simulated flue gas mixtures (ready to use or composed at place), without acid components, are equipped at most with gas dehumidifier (such as bench scale installation), while the pilot plant requires the feed gas conditioning system. This gas treating system is composed of cooling and heat recovery, desulphurization and NO_x removal section, and flue gas drying section. The processed gas is then directed to the carbon dioxide separation section, build of four adsorbers filled with activated carbon, periodically regenerated by reducing the pressure and purge. The applied DR-VPSA technique was used in the pilot plant to CO_2 separation research from real flue gas coming from power plant, and the VPSA technique for CO_2 capture studies from simulated mixture of exhaust gas coming from cement plant.

The developed process configurations have become the subject of patents: PL 231697 (Method of carbon dioxide capture from gas mixtures by vacuum pressure swing adsorption) in the case of bench scale installation and PL237180 (Two-stage method of carbon dioxide capture from exhaust gases using the adsorption method) in the case of pilot plant.

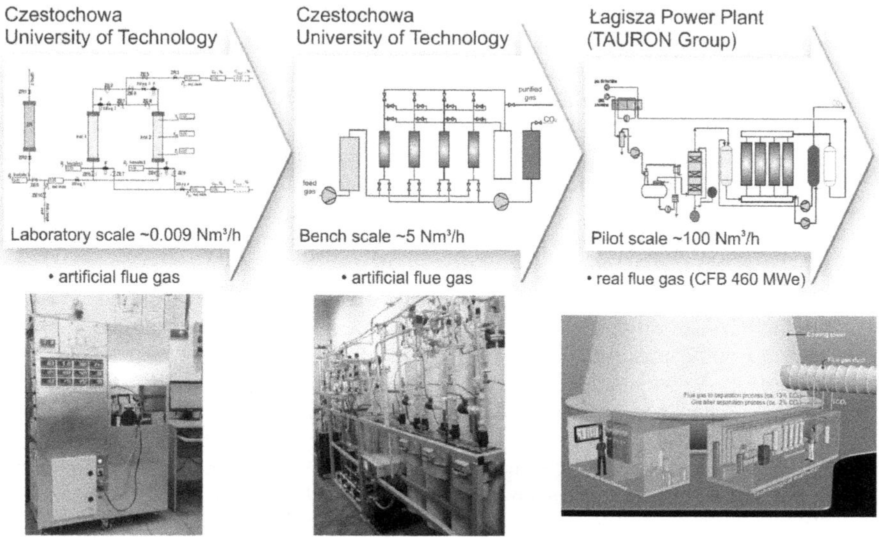

FIGURE 3.1 Scaling of carbon dioxide capture research (reproduced by permission of Publishing Office of Czestochowa University of Technology) [29].

REFERENCES

[1] Dąbrowski, A. Adsorption — from theory to practice, *Advances in Colloid and Interface Science* **2001**, *93*, 135–224, https://doi.org/10.1016/S0001-8686(00)00082-8.

[2] Yang, R. T. *Gas separation by adsorption processes*, Imperial College Press, 1997.

[3] Ruthven, D. N. *Principles of adsorption and adsorption processes*, John Willey & Sons, 1984.

[4] Do, D. D. *Adsorption analysis: equilibria and kinetics*, Imperia College Press, 1998.

[5] Ruthven, D. M.; Farooq, S.; Knaebel, K. S. *Pressure swing adsorption*, VCH Publishers, 1994.

[6] Dhoke, C.; Zaabout, A.; Cloete, S.; Amini, S. Review on reactor configurations for adsorption-based CO_2 capture, *Industrial & Engineering Chemistry Research* **2021**, *60*, 3779–3798, https://doi.org/10.1021/acs.iecr.0c04547.

[7] Sorrels, J. L. Chapter 1 Carbon adsorbers, 2018. Assessed October 15, 2021, https://www.epa.gov/sites/default/files/2018-10/documents/final_carbonadsorberschapter_7thedition.pdf.

[8] Thomas, W. J.; Crittenden, B. *Adsorption technology and design*, 1998, Butterworth-Heinemann.

[9] Suzuki, M. *Adsorption engineering*, 1990, Kodansha.

[10] Choi, S.; Drese, J. H.; Jones, C. W. Adsorbent materials for carbon dioxide capture from large anthropogenic point sources, *ChemSusChem* **2009**, *2*, 796–854, https://doi.org/10.1002/cssc.200900036.

[11] Zeng, Y.; Zou, R.; Zhao, Y. Covalent organic frameworks for CO_2 capture, *Advanced Materials* **2016**, *28*, 2855–2873, https://doi.org/10.1002/adma.201505004.

[12] Thomas-Hillman, I.; Laybourn, A.; Dodds, C.; Kingman, S. W. Realising the environmental benefits of metal–organic frameworks: recent advances in microwave synthesis, *Journal of Materials Chemistry A* **2018**, *6*, 11564–11581, https://doi.org/10.1039/C8TA02919A.

[13] Olajire, A. A. Synthesis chemistry of metal-organic frameworks for CO_2 capture and conversion for sustainable energy future, *Renewable and Sustainable Energy Reviews* **2018**, *92*, 570–607, https://doi.org/10.1016/j.rser.2018.04.073.

[14] Konni, M.; Doddi, S.; Dadhich, A. S.; Mukkamala, S. B. Adsorption of CO_2 by hierarchical structures of f-MWCNTs@Zn/Co-ZIF and N-MWCNTs@Zn/Co-ZIF prepared through in situ growth of ZIFs in CNTs, *Surfaces and Interfaces* **2018**, *12*, 20–25, https://doi.org/10.1016/j.surfin.2018.04.006.

[15] Wu, X.; Shahrak, M. N.; Yuan, B.; Deng, S. Synthesis and characterization of zeolitic imidazolate framework ZIF-7 for CO_2 and CH_4 separation, *Microporous and Mesoporous Materials* **2014**, *190*, 189–196, https://doi.org/10.1016/j.micromeso.2014.02.016.

[16] Chang, P.-H.; Lee, Y.-T.; Peng, C.-H. Synthesis and characterization of hybrid metal zeolitic imidazolate framework membrane for efficient H_2/CO_2 gas separation. *Materials* **2020**, *13*, 5009, https://doi.org/10.3390/ma13215009.

[17] Russell, B. A.; Migone, A. D. Low temperature adsorption study of CO_2 in ZIF-8, *Microporous and Mesoporous Materials* **2017**, *246*, 178–185, https://doi.org/10.1016/j.micromeso.2017.03.030.

[18] Liu, Y.; Liu, J.; Lin, Y. S.; Chang, M. Effects of water vapor and trace gas impurities in flue gas on CO_2/N_2 separation using ZIF-68, *The Journal of Physical Chemistry C* **2014**, *118*, 6744–6751, https://doi.org/10.1021/jp4113969.

[19] Ozdemir, J.; Mosleh, I.; Abolhassani, M.; Greenlee, L. F.; Beitle, R. R.; Beyzavi, M. H. Covalent organic frameworks for the capture, fixation, or reduction of CO_2, *Frontiers in Energy Research* **2019**, *7*, 77, https://doi.org/10.3389/fenrg.2019.00077.

[20] Kundu, N.; Sarkar, S. Porous organic frameworks for carbon dioxide capture and storage, *Journal of Environmental Chemical Engineering* **2021**, *9*, 105090, https://doi.org/10.1016/j.jece.2021.105090.

[21] García, S.; Gil, M. V.; Martín, C. F.; Pis, J. J.; Rubiera, F.; Pevida, C. Breakthrough adsorption study of a commercial activated carbon for pre-combustion CO_2 capture, *Chemical Engineering Journal* **2011**, *71*, 549–556, https://doi.org/10.1016/j.cej.2011.04.027.

[22] Majchrzak-Kucęba, I.; Wawrzyńczak, D.; Ściubidło, A. Application of metal-organic frameworks in VPSA technology for CO_2 capture, *Fuel* **2019**, *255*, 115773, https://doi.org/10.1016/j.fuel.2019.115773.

[23] Liu, Z.; Grande, C. A.; Li, P.; Yu, J. Rodrigues, A. E., Adsorption and desorption of carbon dioxide and nitrogen on zeolite 5A, *Separation Science and Technology* **2011**, *46*, 434–451, https://doi.org/10.1080/01496395.2010.513360.

[24] Suzuki, T.; Sakoda, A.; Suzuki, M.; Izumi, J. Recovery of carbon dioxide from stack gas by piston-driven ultra-rapid PSA, *Journal of Chemical Engineering of Japan* **1997**, *30*, 1026–1033, https://doi.org/10.1252/jcej.30.1026.

[25] Cho, S.-H.; Park, J.-H.; Beum, H.-T.; Han, S.-S.; Kim, J.-N. A 2-stage PSA process for the recovery of CO_2 from flue gas and its power consumption. In *Studies in surface science and catalysis, proceedings of the 7th international conference on carbon dioxide utilization*; Park S.-E., Chang J.-S., Lee K.-W., Eds., 2014, *153*, 405–410, https://doi.org/10.1016/S0167-2991(04)80287-8.

[26] Wang, L.; Yang, Y.; Shen, W.; Kong, X.; Li, P.; Yu, J.; Rodrigues, A. E. CO_2 capture from flue gas in an existing coal-fired power plant by two successive pilot-scale VPSA units, *Industrial and Engineering Chemistry Research* **2013**, *52*, 7947–7955, https://doi.org/10.1021/ie4009716.

[27] Wawrzyńczak, D.; Majchrzak-Kucęba, I.; Srokosz, K.; Kozak, M.; Nowak, W.; Zdeb, J.; Smółka, W.; Zajchowski, W. The pilot dual-reflux vacuum pressure swing adsorption unit for CO_2 capture from flue gas, *Separation and Purification Technology* **2019**, *209*, 560–570, https://doi.org/10.1016/j.seppur.2018.07.079.

[28] Mofarahi, M.; Khojasteh, Y.; Khaledi, H.; Farahnak, A. Design of CO_2 absorption plant for recovery of CO_2 from flue gases of gas turbine, *Energy* **2008**, *33*, 1311–1319, https://doi.org/10.1016/j.energy.2008.02.013.
[29] Wawrzyńczak, D.; Majchrzak-Kucęba, I.; Nowak, W.; Srokosz, K.; Kozak, M. Wychwyt CO_2 metodą adsorpcyjną – doświadczenia z pilotowej instalacji DR-VPSA. In *Spalanie tlenowe dla kotłów pyłowych i fluidalnych zintegrowanych z wychwytem CO_2. Doświadczenia z instalacji pilotowych i perspektywy dla instalacji demonstracyjnych*; Nowak, W., Ściązko, M., Czakiert, T., Eds.; Wydawnictwo Politechniki Częstochowskiej, 2015, 184–195.
[30] Voss, C. Applications of pressure swing adsorption technology, *Adsorption* **2005**, *11*, 527–529, https://doi.org/10.1007/s10450-005-5979-3.
[31] Na, B.-K.; Koo, K.-K.; Eum, H.-M.; Lee, H.; Song, H. K. CO_2 recovery from flue gas by PSA process using activated carbon, *Korean Journal of Chemical Engineering* **2001**, *18*, 220–227, https://doi.org/10.1007/BF02698463.
[32] Shen, C.; Yu, J.; Li, P.; Grande, C. A. Rodrigues, A. E., Capture of CO_2 from flue gas by vacuum pressure swing adsorption using activated carbon beads, *Adsorption* **2011**, *17*, 179–188, https://doi.org/10.1007/s10450-010-9298-y.
[33] Shen, C.; Liu, Z.; Li, P.; Yu, J. Two-stage VPSA process for CO_2 capture from flue gas using activated carbon beads, *Industrial & Engineering Chemistry Research* **2012**, *51*, 5011–5021, https://doi.org/10.1021/ie202097y.
[34] Webley, P. A.; Qader, A.; Ntiamoah, A.; Ling, J.; Xiao, P.; Zhai, Y. A new multi-bed vacuum swing adsorption cycle for CO_2 capture from flue gas streams, *Energy Procedia* **2017**, *114*, 2467–2480, https://doi.org/10.1016/j.egypro.2017.03.1398.
[35] Ntiamoah, A.; Ling, J.; Xiao, P.; Webley, P. A.; Zhai, Y. CO_2 capture by vacuum swing adsorption: role of multiple pressure equalization steps, *Adsorption* **2015**, *21*, 509–522, https://doi.org/10.1007/s10450-015-9690-8.
[36] Wang, L.; Liu, Z.; Li, P.; Yu, J.; Rodrigues, A. E. Experimental and modelling investigation on post-combustion carbon dioxide capture using zeolite 13X-APG by hybrid VTSA process, *Chemical Engineering Journal* **2012**, *197*, 151–161, https://doi.org/10.1016/j.cej.2012.05.017.
[37] Grande, C. A.; Cavenati, S.; Rodrigues, A. E. Pressure swing adsorption for carbon dioxide sequestration. In *2nd Mercosur congress on chemical engineering and 4th Mercosur congress on process systems engineering (ENPROMER2005)*, 2005, 1–10, Rio de Janeiro, Brazil.
[38] Wawrzyńczak, D.; Bukalak, D.; Majchrzak-Kucęba, I.; Nowak, W. Effect of desorption pressure on CO_2 separation from combustion gas by means of zeolite 13X and activated carbon, *Polish Journal of Environmental Studies* **2014**, *23*, 1437–1440, http://pjoes.com/pdf-89334-23191?filename=Effect%20of%20Desorption.pdf.
[39] Plaza, M. G.; García, S.; Rubiera, F.; Pis, J. J.; Pevida, C. Post-combustion CO_2 capture with a commercial activated carbon: comparison of different regeneration strategies, *Chemical Engineering Journal* **2010**, *163*, 41–47, https://doi.org/10.1016/j.cej.2010.07.030.
[40] Wawrzyńczak, D. Adsorpcja CO_2 pochodzącego ze spalania węgla w atmosferze wzbogaconej tlenem w procesie zmiennociśnieniowym i zmiennotemperaturowym: PhD thesis, 2012, Częstochowa.
[41] Merel, J.; Clausse, M.; Meunier, F. Experimental investigation on CO_2 post–combustion capture by indirect thermal swing adsorption using 13X and 5A zeolites, *Industrial & Engineering Chemistry Research* **2008**, *47*, 209–215, https://doi.org/10.1021/ie071012x.
[42] Zhang, J.; Webley, P. A.; Xiao, P. Effect of process parameters on power requirements of vacuum swing adsorption technology for CO_2 capture from flue gas, *Energy Conversion and Management* **2008**, *49*, 346–356, https://doi.org/10.1016/j.enconman.2007.06.007.

[43] Chaffee, A. L.; Knowles, G. P.; Liang, Z.; Zhang, J.; Xiao, P.; Webley, P. A. CO_2 capture by adsorption: materials and process development, *International Journal of Greenhouse Gas Control* **2007**, *1*, 11–18, https://doi.org/10.1016/S1750-5836(07)00031-X.

[44] Ishibashi, M.; Ota, H.; Akutsu, N.; Umeda, S.; Tajika, M.; Izumi, J.; Yasutake, A.; Kabata, T.; Kageyama, Y. Technology for removing carbon dioxide from power plant flue gas by the physical adsorption method, *Energy Conversion and Management* **1996**, *37*, 929–933, https://doi.org/10.1016/0196-8904(95)00279-0.

[45] Ishibashi, M.; Otake, K.; Kanamori, S.; Yasutake, A. Study on CO_2 removal technology from flue gas of thermal power plant by physical adsorption method. In *Proceedings of the fifth international conference on greenhouse gas control technologies*; Williams, D. J., Durie, R. A., McMullan, P., Paulson, C. A. J., Smith, A. Y., Eds.; CSIRO Publishing, 2001, Australia, 95–106.

[46] Takamura, Y.; Narita, S.; Aoki, J.; Hironaka, S.; Uchida, S. Evaluation of dual-bed pressure swing adsorption for CO_2 recovery from boiler exhaust gas, *Separation and Purification Technology* **2001**, *24*, 519–528, https://doi.org/10.1016/S1383-5866(01)00151-4.

[47] Liu, Z.; Wang, L.; Kong, X.; Li, P.; Yu, J.; Rodrigues, A. E. Onsite CO_2 capture from flue gas by an adsorption process in a coal-fired power plant, *Industrial and Engineering Chemistry Research* **2012**, *51*, 7355–7363, https://doi.org/10.1021/ie3005308.

[48] Wang, L.; Yang, Y.; Shen, W.; Kong, X.; Li, P.; Yu, J.; Rodrigues, A. E. Experimental evaluation of adsorption technology for CO_2 capture from flue gas in an existing coal-fired power plant, *Chemical Engineering Science* **2013**, *101*, 615–619, https://doi.org/10.1016/j.ces.2013.07.028.

[49] Krishnamurthy, S.; Rao, V. R.; Guntuka, S.; Sharratt, P.; Haghpanah, R.; Rajendran, A.; Amanullah, M.; Karimi, I. A.; Farooq, S. CO_2 capture from dry flue gas by vacuum swing adsorption: a pilot plant study, *American Institute of Chemical Engineers Journal* **2014**, *60*, 1830–1842, https://doi.org/10.1002/aic.14435.

[50] Webley, P. A. Adsorption technology for CO_2 separation and capture: a perspective, *Adsorption* **2014**, *20*, 225–231, https://doi.org/10.1007/s10450-014-9603-2.

[51] Majchrzak-Kucęba, I.; Wawrzyńczak, D.; Ściubidło, A.; Zdeb, J.; Smółka, W.; Zajchowski, A. Assessment of the sorption capacity of activated carbon used in desulphurization and NO_x removal section in pilot DR-VPSA CO_2 capture unit. In *Energy systems in the near future: energy, exergy, ecology and economics*; Stanek, W., Czarnowska, L., Kostowski, W., Gładysz, P., Eds.; Gliwice, 2018, 891–899.

4 Bio-based adsorbents for CO_2 separation

Nausika Querejeta, María Victoria Gil, María Rodríguez, Fernando Rubiera, and Covadonga Pevida

CONTENTS

DOI: 10.1201/9781003336587-4

4.1 INTRODUCTION

Biomass is a raw material of renewable origin and geographically distributed worldwide. Apart from energy recovery through various processes, another possibility is to reuse the residues from the exploitation or transformation of biomass, such as the development of adsorbent materials.

The common sources of biomass are agricultural, forest, municipal, energy, and biological. Likewise, classification of the type of biomass feedstock in wet (moisture content >30%) or dry (<30%) is of vital importance for the selection of the pre-treatment method and assessment of the feasibility of the production of biochar, hydrochar, and activated carbon (AC). Biomass typically consists of three main components: cellulose, hemicellulose, and lignin, which exist in many different proportions depending on the type of biomass. Cellulose is the primary, strong crystalline structural component of cell walls in biomass; hemicellulose has a random amorphous structure with little strength which composition and structure vary significantly among different biomasses; and finally lignin is an integral part of the secondary cell walls of plants and the cementing agent for cellulose fibers holding adjacent cells together [1].

4.2 ADSORBENTS PRODUCTION TECHNOLOGIES

Currently, several techniques are available for the production of carbonaceous adsorbents from biomass. However, depending upon the type of feedstock (wet or dry) and the desired properties of bio-based adsorbents for CO_2 capture applications, the choice of pre-treatment method(s) is very limited.

4.2.1 Pyrolysis

Carbonization is a slow pyrolysis process, in which the production of a carbon-rich solid product (biochar) is the primary goal. The biomass is heated slowly (10–30°C/min) in the absence of oxygen within a temperature range of 300–650°C and a long residence time (few minutes to a couple of hours) [1, 2].

During slow pyrolysis, cellulose, hemicellulose, and lignin break down into smaller hydrocarbon molecules, producing some condensable gases, which may decompose further into non-condensable gases (CO, CO_2, H_2, and CH_4) and liquid (bio-oil); and biochar. This decomposition occurs partly through gas-phase homogeneous reactions and partly through gas-solid-phase heterogeneous thermal reactions [1].

4.2.2 Hydrothermal carbonization

Hydrothermal carbonization (HTC) applied to biomass precursors results simple, low cost, and energy efficient. It does not require the use of organic solvents, catalysts, or surfactants, so it can be classified as "green" [3].

During HTC, cellulose, hemicellulose, and lignin decompose following a complex cascade of aldol reactions, cycloadditions, and condensations to obtain a solid material rich in carbon (hydrochar). The use of subcritical water in the process results in less non-condensable products, increasing the reaction yield.

This process is energetically favorable when compared to traditional pyrolysis because it uses milder conditions, does not need the previous drying of the precursors, and is exothermic; in fact, the heat released during the reactions accounts for 33% of the energy required to complete the process [4]. The hydrochars also have good self-binding properties, which is very interesting for their subsequent conformation into pellets [4, 5].

4.2.3 Activation

Activation is a process mainly used for further increasing the porosity of any char or porous carbon. In this sense, the application of the biochar/hydrochar as a low-cost adsorbent material in gas separations often requires additional activation steps to increase the surface area and pore volume.

Physical and chemical activation methods are the two common techniques for the activation of carbonaceous materials. Physical activation is a process of selective gasification of carbon atoms and involves prior carbonization of the carbon precursor in an inert atmosphere followed by the heat treatment of the char in the presence of a gasifying agent [6]. During carbonization, most of the heteroatoms in the carbon matrix precursor, such as O, H, and N, evolve with volatiles, which results in the enrichment of the carbon content, increase in aromaticity, and incipient microporosity development. During the activation stage, the reaction between the carbon atoms and the gas used as the activating agent (CO_2, NH_3, air, steam, etc.) occurs. In chemical activation, the carbon precursor is mixed/impregnated with a chemical activating agent such as KOH, NaOH, NH_3, K_2CO_3, and $ZnCl_2$, and acids such as H_3PO_4 and H_2SO_4, and subsequently carbonized and activated at a temperature ranging from 400 to 900°C under an inert atmosphere (usually N_2 or Ar). The latter needs lower temperature and shorter activation time, which often renders a higher specific surface area and more uniform pore size distribution (PSD).

4.3 POROUS ADSORBENT MATERIALS FOR CO_2 CAPTURE

The application of carbon-based adsorbents to CO_2 capture has been extensively studied. Carbon-based adsorbents promising attributes include low cost, high surface area, high amenability to pore structure modification and surface functionalization, and relative ease of regeneration [7]. Furthermore, owing to their hydrophobic nature, carbon-based adsorbents show high stability in humid conditions, though a decrease in capacity is often observed compared to the performance under dry

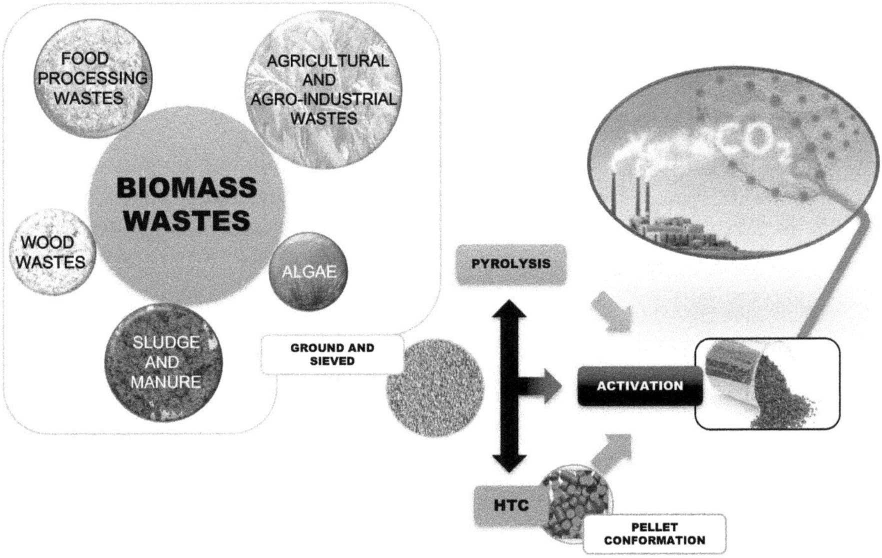

FIGURE 4.1 Schematic diagram representing the production of biochar/hydrochar and activated carbon from biomass for post-combustion CO_2 capture applications.

conditions [8]. All of these features make them appealing adsorbent candidates for post-combustion CO_2 capture applications.

Among the carbon-based adsorbents investigated so far, focus should be on biochars, hydrochars, and ACs. Biochar is the solid product from dry carbonization or pyrolysis [9], hydrochar is generated as a slurry (a two-phase mixture of solid and liquid) via HTC for subsequent thermochemical conversion [3], and AC can be obtained by physical activation (generally in two stages) and chemical activation (see Figure 4.1).

4.4 BIO-BASED MATERIALS AS POTENTIAL CO_2 ADSORBENTS

Any carbonaceous material with high carbon and low ash contents could be used as a precursor for the production of ACs, being preferably available, cheap, and renewable [10]. With this regard, the recycling of biomass wastes has received increasing attention in recent years [11]. Their use as precursors in the production of porous carbons offers benefits to reduce the total cost of capture and the environmental impact resulting from the uncontrolled disposal of residues [12].

At the end of this section, Table 4.1 collects relevant data on the CO_2 capture performance and characteristics of CO_2 bio-adsorbents.

4.4.1 Agricultural and Agro-Industrial Wastes

Crop residues encompass all agricultural wastes such as African palm stones, almond shells, *Cannabis sativa L.*, cherry stones, coconuts shells, coffee grounds, olive stones, peanut shells, pine nut shells, rice husks, sugarcane bagasse, and walnut shells.

4.4.1.1 African palm stones

Vargas et al. [13] prepared a series of binderless AC carbon monoliths by H_3PO_4 chemical activation of African palm stones evaluating several concentrations of phosphoric acid. The AC monoliths exhibited an improved CO_2 adsorption behavior with the increase in the activation degree, reaching a maximum value of approximately 3.7 mol/kg at 0°C for sample PACM-28.

4.4.1.2 Almond shells

Plaza et al. [14] subjected almonds shells to activation with CO_2 and heat treatment in ammonia gas to develop carbon-based adsorbents. The CO_2 adsorption capacity for the activated adsorbents with CO_2 reached 2.7 mol/kg in pure CO_2 flow and 1.2 mol/kg in a binary mixture of 15% CO_2 in N_2, both at 25°C. Regarding the ammonia-treated carbons, the capacities outcome 2.2 and 1.1 mol/kg, respectively. Likewise, González et al. [15] produced sustainable carbon adsorbents from almond shells by single-step activation with CO_2. Their CO_2 adsorption capacity measured at low pressures was between 0.6 and 1.1 mol/kg at 15 kPa and 25–50°C.

4.4.1.3 Cannabis

Yang et al. [16] prepared a series of ACs by chemical activation of hemp (*Cannabis sativa L.*) stem with KOH. The AC prepared at 800°C produced a good BET surface area of 3241×10^3 m^2/kg and a total pore volume of 1.98×10^{-3} m^3/kg. The increase in heat treatment temperature beyond 600°C decreased the ultramicropore volume with the consequent reduction in the CO_2 uptake.

4.4.1.4 Cherry stones

Álvarez-Gutierrez et al. [17] optimized the production of two cherry stones-based adsorbents by single-step activation with CO_2 (CS-CO_2) and steam (CS-H_2O). The activation conditions that maximized the CO_2 adsorption capacity by the adsorbents at 25°C and atmospheric pressure were determined by response surface methodology (RSM). The highest uptakes were 2.6 mol/kg for CS-CO_2 and 2.4 mol/kg for CS-H_2O.

4.4.1.5 Coconut shells

Yang et al. [18] prepared three kinds of ACs using coconut shells as precursors. The highest CO_2 adsorption capacity was 2.6 mol/kg at 0.2 MPa and 25°C for the sample impregnated into a phosphoric acid solution and carbonized at 600°C for 2 h. Ello et al. [19] developed an AC from coconut shells by one-step activation with CO_2 at 800°C. The highest CO_2 uptake was reached at atmospheric pressure, with values of 3.9 mol/kg at 25°C and 5.6 mol/kg at 0°C. Likewise, Abdeljaoued et al. [20] prepared a coconut shells-based AC by using CO_2 one-pot activation procedure. The adsorption capacity for CO_2 was 1.9 mol/kg at 30°C and 101 kPa. Vargas et al. [13] also developed a series of AC monoliths by chemical activation with H_3PO_4 of coconut shells. Concerning the adsorption of CO_2 at atmospheric pressure and 0°C, AC monoliths reached an optimum value of ~3.7 mol/kg for sample CACM32.

4.4.1.6 Coffee grounds

Plaza et al. [21] studied the valorization of spent coffee grounds as CO_2 adsorbents for post-combustion capture applications. Two different activation methods were compared: CO_2 physical activation and KOH chemical activation. The obtained ACs presented CO_2 adsorption capacities up to 4.8 mol/kg at 0°C and 3.0 mol/kg at 25°C. In addition, Plaza et al. [22] prepared a low-cost adsorbent by physical CO_2 activation from spent coffee grounds. The CO_2 adsorption capacity resulted in 0.8 and 1.2 mol/kg at 50°C and 25°C, respectively, under a pressure of 15 kPa. More recently, Querejeta et al. [23] optimized the preparation of high-performance post-combustion CO_2 capture adsorbents using HTC and subsequent CO_2 activation from spent coffee grounds. The maximum value of CO_2 capture capacity at 50°C and a CO_2 partial pressure of 10 kPa, 0.7 mol/kg, were attained at both the lowest temperature and dwell time (120°C, 3 h).

4.4.1.7 Olive stones

Plaza et al. [24] studied two approaches for adsorbent production from olive stones with CO_2 separation purposes: activation with CO_2 and heat treatment with gaseous ammonia. Adsorbents activated with CO_2 presented a CO_2 capture capacity of 2.4 mol/kg at 25°C and 0.7 mol/kg at 100°C. Alternatively, the aminated samples reached the maximum values when produced at 800°C, with a CO_2 uptake of 2.0 mol/kg at 25°C and 0.6 mol/kg at 100°C. Besides, González et al. [15] prepared porous carbon adsorbents for CO_2 capture by one-pot activation of olive stones using CO_2 as activating agent. The olive stones-based carbons showed a CO_2 uptake up to 3.1 mol/kg at 120 kPa and 25°C, and 0.6–1.1 mol/kg at 15 kPa and 25–50°C. More recently, Puig-Gamero et al. [25] produced ACs from olive stones by single-step activation with steam and CO_2, in a bench-scale high-pressure thermobalance. The ACs showed a CO_2 adsorption capacity of 4.3 and 4.7 mol/kg at 30°C and 1000 kPa for activation with CO_2 and steam, respectively.

4.4.1.8 Peanut shells

Recently, Sher et al. [26] prepared biomass-based ACs using a single-step chemical activation process with different KOH/C mass ratios. The CO_2 adsorption measurements of ACs samples at 25°C and 15 kPa revealed a CO_2 adsorption peak of 3.9 mol/kg while the overall CO_2 adsorption is equivalent to 80% in the initial 180 to 240 s. After this initial 80% adsorption capacity, horizontal plateaus of adsorption capacity were observed for samples named PNK1 and PNK2 indicating the saturation of available micropores. A constant increase in adsorption with time was observed for the PNK3 sample, whereas for PNK4 a decrease in adsorption capacity was observed.

4.4.1.9 Pine nut shells

Deng et al. [27] used pine nut shells to prepare ACs for the adsorption of CO_2 with different KOH/C mass ratios at different activation temperatures. The highest CO_2 uptakes were reported at a KOH/C ratio of 2 and activation temperature of 700°C:

5.0 mol/kg at 25°C and 101 kPa and 7.7 mol/kg at 0°C and 101 kPa. Under typical flue gas conditions (100°C and 15 kPa), micropores in the range of 0.22 and 0.40 nm were the only effective, with a CO_2 uptake of 0.7 mol/kg that increased when reducing the temperature.

4.4.1.10 Rice husks

Li et al. [28] prepared porous carbons for CO_2 capture at low pressures by KOH activation of rice husk char. The ACs exhibited a high CO_2 uptake, 2.1 mol/kg and 6.2 mol/kg, at 0°C for 10 kPa and 101 kPa, respectively.

4.4.1.11 Sugarcane bagasse

Creamer et al. [29] produced biochars from sugarcane bagasse (BG) through slow pyrolysis in a N_2 environment at three different temperatures. The highest CO_2 adsorption capacity at 25°C, 1.7 mol/kg, was observed on the BG produced at 600°C.

4.4.1.12 Walnut shells

Chomiak et al. [30] optimized the properties of walnut shells-based ACs for CO_2 adsorption. They prepared a series of microporous granular ACs (GACs) by KOH activation. Suitable heat temperature treatment allowed preparing adsorbents with high CO_2 uptake at 0°C of 7.2 and 18.2 mol/kg under 101 kPa and 3 MPa, respectively. At 25°C, the values reached 3.0 and 12.6 mol/kg. Likewise, Asadi-Sangachini et al. [31] prepared ACs from walnut shells by chemical activation with two different activating agents (KOH and H_3PO_4) and using different impregnation ratios. The highest CO_2 adsorption capacities, 2.8 mol/kg and 3.6 mol/kg, were obtained at 101 kPa and 30°C for the activation with KOH at an impregnation ratio of 1.5 and for H_3PO_4 at a ratio of 2.5, respectively. In addition to peanut shells, Sher et al. [26] also studied walnut shells as precursors for post-combustion CO_2 capture adsorbents. The maximum CO_2 uptake, 3.2 mol/kg, was obtained at 25°C and atmospheric pressure for a walnut shells-derived AC prepared with a KOH/C of 3 and activated at 750°C.

4.4.2 Wood wastes

Wood wastes generally are concentrated at processing factories, i.e., plywood mills and sawmills.

4.4.2.1 Bamboo

Wang et al. [32] prepared an AC with a high specific surface area and considerable mesopores from bamboo scraps by phosphoric acid activation. The AC1 adsorbed much more CO_2 at 25°C than CH_4, N_2, and O_2, and thus became a good adsorbent for separating CO_2 from its mixture with nitrogen, oxygen, or both. Likewise, Wei et al. [33], used bamboo as a precursor for AC preparation through KOH activation. The AC showed high CO_2 capacity and good reversibility of the adsorption after activation at 700°C and 1.5 h of activation time. The CO_2 uptake determined was 7.0 mol/kg at 0°C and 101 kPa. More recently, Khuong et al. [34] converted bamboo and its solid residue after hydrothermal treatment into a porous carbon by CO_2 activation.

The highest adsorption capacity of 3.4 mol/kg at 25°C and 101 kPa was obtained for the hydrochar activated at 800°C for 900 s.

4.4.2.2 Eucalyptus

Heidari et al. [35] prepared a series of ACs from *Eucalyptus camaldulensis* wood by chemical activation with H_3PO_4 and $ZnCl_2$ at different impregnation ratios, as well as by pyrolysis followed by activation with KOH. The highest CO_2 adsorption capacity was 4.1 mol/kg at 101 kPa and 30°C for the carbon activated with KOH at 900°C.

4.4.2.3 Hickory wood

In addition to sugarcane bagasse, Creamer et al. [29] produced biochars from hickory wood (HW) through slow pyrolysis in N_2 at three different temperatures. The highest CO_2 adsorption capacity was 1.4 mol/kg at 25°C and 101 kPa.

4.4.2.4 Leucaena

Huang et al. [36] carried out microwave co-torrefaction of sewage sludge and Leucaena wood to produce biochar for the adsorption of CO_2. The adsorption capacity of pure Leucaena wood biochar was approximately 1.2 mol/kg, almost four times higher than that of pure sewage sludge biochar. The highest fraction of Leucaena wood in the blend resulted in lower CO_2 adsorption capacity.

4.4.2.5 Mesquite

Li et al. [37] used mesquite biochar as the feedstock for a one-step KOH activation to obtain high surface area CO_2 sorbents. The transformation of mesquite wood into a carbon-rich material in a single-step KOH activation at 800°C resulted in a CO_2 uptake of 26.0 mol/kg at 3 MPa and 25°C.

4.4.2.6 Pine

Alongside walnut and peanut shells, Sher et al. [26] evaluated the CO_2 uptake for pine wood-ACs produced by single-step chemical activation with different KOH/C mass ratios. The highest adsorption capacity, 3.5 mol/kg at 15 kPa, was attained by the carbon activated (KOH/C ratio of 1) at 750°C for 1 h.

4.4.2.7 Sawdust

Zhu et al. [38] investigated the effect of the synthesis conditions on the CO_2 capture capacity of the ACs obtained by single-step KOH activation of Paulownia sawdust. AS-4-700-1, which was prepared with the optimized KOH/C mass ratio of 4, activation temperature of 700°C and activation time of 1 h, exhibited the highest CO_2 capture capacity of 8.0 mol/kg at 0°C and 101 kPa. In an attempt to overcome the main drawbacks of developing carbon adsorbents from pine sawdust, such as low carbon yield and poor mechanical properties of the resulting carbons, Durán et al. [39] assessed the effect of the addition of coal tar pitch. Adsorbent pellets were produced from pine sawdust and coal tar pitch by activation with CO_2. For an activation time of 2 h at a temperature of 800°C and using pellets of a blend of pine sawdust and coal tar pitch in a mass ratio of 10:5 (PPSCT105) pre-oxidized with air at 300°C for 1 h, an adsorption capacity of 0.5 mol/kg in a mixture with 10 vol% CO_2 (balance N_2) at atmospheric pressure and 50°C was achieved.

4.4.3 Algae

The term algae refers to a diverse group of highly productive organisms that include microalgae, macroalgae (seaweed), and cyanobacteria (formerly called "blue-green algae").

4.4.3.1 Macroalgae

Tian et al. [40] developed hierarchical porous carbons from the direct carbonization of *Enteromorpha prolifera* algae after a freeze-drying treatment. The CO_2 uptake was 2.0 mol/kg at 101 kPa and 6.5 mol/kg at 2 MPa and 25°C for the sample carbonized at 900°C for 1 h under N_2 atmosphere (EPC-900).

4.4.3.2 Microalgae

Sevilla et al. [41] developed highly microporous N-doped carbons by KOH chemical activation of a hydrochar obtained from a mixture of microalgae (*Spirulina platensis*) and glucose. The highest CO_2 adsorption capacity was 7.4 mol/kg at 0°C and 101 kPa for the HTC carbon with a KOH/HTC ratio of 4 and activated at 700°C. Furthermore, Durán et al. [42] produced ACs using different microalgae species (*Chlorella Spirulina*, *Acutodesmus Obliquus*, and *Coelastrella sp.*) as well as mixtures of pine sawdust and microalgae. The *Spirulina* mixed with pine sawdust and directly activated with CO_2 presented an outstanding CO_2 adsorption capacity of 0.9 mol/kg at 50°C and a CO_2 partial pressure of 10.5 kPa.

4.5 BIO-ADSORBENTS IN CO_2 CAPTURE APPLICATIONS

Most of the literature on bio-based adsorbents for CO_2 capture relies on equilibrium of adsorption data from isotherms or thermogravimetric analysis at different temperatures and CO_2 partial pressures. However, the analysis of the bio-based adsorbent performance in a more realistic scenario requires dynamic CO_2 adsorption studies at conditions representative of real gas streams, and hence cyclic breakthrough experiments better reproduce practical separations. Besides, decision-making of a bio-adsorbent for a particular CO_2 capture application requires systematic information on the feedstock (cost, characteristics, or availability), physicochemical properties of the bio-adsorbent, regenerability, and impact of the presence of multiple gas components, alongside relevant cost-benefit and environmental performance.

The bio-adsorbents must exhibit relatively fast adsorption and desorption kinetics for their successful implementation in swing adsorption processes. In this regard, the number of studies available is relatively modest compared to the large amount of research already conducted on developing ACs from biomass precursors. This section will review studies that go beyond the milligram scale, from breakthrough tests in laboratory fixed-bed setups to the scarce number of initiatives at a larger scale.

TABLE 4.1
CO_2 adsorption performance and characteristics of non-functionalized bio-based adsorbents

			S_{BET}	V_P	V_{DR}	W_{0,CO_2}	Temperature	CO_2 partial pressure	CO_2 uptake	
Feedstock	Carbonization	Activation	m^2/kg 10^3	m^3/kg 10^{-3}	m^3/kg 10^{-3}	m^3/kg 10^{-3}	°C	kPa	mol/kg	Ref.
African palm stones		H_3PO_4	1092	0.46	0.42	0.36	0	101	3.7	[8]
Almond shells	N_2 at 600°C	CO_2 at 700°C	1090	0.50	0.42	0.12	25	101	2.7	[9]
Almond shells		CO_2 at 750°C	822	0.37	0.31		25	101	2.7	[10]
Cannabis sativa L.	N_2 at 500°C	KOH	1365	0.73		0.54	0	101	7.8	[11]
Cherry stones		CO_2 at 885°C	1045	0.48	0.4	0.35	25	101	2.6	[12]
Coconut shells		H_3PO_4	1922	N/A	0.68	N/A	25	200	2.6	[13]
Coconut shells		CO_2 at 800°C	1327	0.65	0.55	N/A	25	101	3.9	[14]
Coconut shells		CO_2 at 900°C	1378	0.63	0.54	0.47	30	101	1.9	[15]
Coconut shells		H_3PO_4	1322	0.61	0.49	0.42	30	101	3.7	[8]
Coffee grounds	N_2 at 400°C	KOH	840	0.37	0.34	0.36	25	101	3.0	[16]
Coffee grounds		CO_2 at 700°C	522	0.21	0.21	0.23	25	15	0.8	[17]
Coffee grounds	HTC at 120°C	CO_2 at 800°C	801	0.34	0.30	0.29	50	10	0.7	[18]
Olive stones	N_2 at 600°C	CO_2 at 800°C	1079	0.50	N/A	0.25	25	101	2.4	[19]
Olive stones		CO_2 at 800°C	1215	0.51	0.48	N/A	25	120	3.1	[10]
Olive stones		CO_2 at 1000°C	955	0.44	0.40	N/A	30	1000	4.7	[20]
Macadamia nut shells	N_2 at 700°C	CO_2 at 900°C	512	N/A	N/A	0.18	25	13	1.1	[38]
Peanut shells		KOH	901	0.38	0.33	N/A	25	15	3.9	[21]
Pine nut shells	N_2 at 600°C	KOH	N/A	N/A	N/A	N/A	25	15	2.0	[22]
Rice husks		KOH	774	0.41	0.30	0.15	0	10	2.1	[23]
Sugarcane bagasse	N_2 at 600°C		388	N/A	N/A	N/A	25	101	1.7	[24]

Wallnut shells	N_2 at 700°C	KOH	1982	0.82	0.80	0.64	0	3000	18.2	[25]
Wallnut shells	N_2 at 600°C	H_3PO_4	1513	1.42	1.06	N/A	30	101	3.6	[26]
Wallnut shells		KOH	147	0.26	0.04	N/A	25	101	3.2	[21]
Winery wastes	HTC at 200°C	KOH	2029	0.85	0.81	0.66	25	2000	15.7	[39]
Pig manure	N_2 at 500°C		N/A	0.04	N/A	N/A	25	101	0.5	[40]
Horse manure (hydrochar)		CO_2 at 800°C	749	0.82	0.14	0.13	0	10	1.4	[41]
Sewage sludge	N_2 at 700°C	NaOH	179	0.25	0.01	N/A	25	101	1.3	[42]
Bamboo scraps		H_3PO_4	1567	1.42	0.75	N/A	25	2500	11.5	[27]
Bamboo		KOH	1846	N/A	0.78	0.36	25	101	4.5	[28]
Moso bamboo	HTC at 200°C	CO_2 at 800°C	952	0.40	0.80	N/A	25	101	3.4	[29]
Eucalyptus		K_2CO_3 and KOH	2595	1.28	1.24	N/A	30	101	4.1	[30]
Hickory wood	N_2 at 600°C		401	N/A	N/A	N/A	25	101	1.4	[24]
Mesquite	N_2 at 450°C	KOH	3167	1.65	N/A	N/A	25	3000	26.0	[32]
Pine wood		KOH	944	0.35	0.33	N/A	25	101	3.5	[21]
Paulownia sawdust		KOH	1643	0.86	0.63	N/A	0	101	8.0	[33]
Pine sawdust	Pre-O_2 at 250°C	CO_2 at 800°C	783	0.33	0.30	N/A	50	10	0.5	[34]
Spirulina platensis/glucose	HTC at 200°C	KOH	2390	1.15	0.96	0.34	0	101	7.4	[36]
Spirulina/pine wood		CO_2 at 800°C					50	10.5	0.9	[37]

4.5.1 Post-combustion CO_2 capture

Most of the studies published in the literature of bio-based CO_2 adsorbents have focused on post-combustion CO_2 capture applications due to the potential use of bio-wastes that inherently bring low or even no-cost for the feedstock.

Our research group has been very intensively evaluating the dynamic CO_2 adsorption performance of in-house produced bio-based adsorbents in a broad range of CO_2 separation applications with a particular emphasis in post-combustion capture. González et al. [15] evaluated olive stone and almond shells-derived ACs produced by single-step activation with CO_2 at 800°C, in a binary mixture of CO_2/N_2 (14/86 vol%) at 50°C and 120 kPa, reporting CO_2 adsorption capacities of around 0.6 mol/kg and apparent CO_2/N_2 selectivity in the range of 20–30 in a fixed-bed laboratory setup. These selectivity values exceeded those estimated from the equilibrium adsorption isotherms pointing to the CO_2 displacing the weaker adsorbed N_2, a phenomenon not shown in the static experiments. Our group also conducted breakthrough (BT) experiments and Vacuum Swing Adsorption (VSA) cycles on a spent coffee grounds-based AC conformed into pellets (patent filed ESES252625) at 50°C and 130 kPa and feed concentrations of CO_2 between 9 and 31 vol% (N_2 balance) [22] and corroborated the displacement of adsorbed N_2 by the preferential adsorption of the slower moving CO_2. The average CO_2 adsorption capacities during the BT experiment ranged between 0.5 and 1.1 mol/kg with the increased CO_2 concentration in the feed. The VSA experiments showed that a feed with 14 vol% CO_2 could be concentrated up to 75% (recovering 84% of the CO_2 in the feed) in a VTSA configuration consisting of pressurization with feed, adsorption at 50°C and 130 kPa, rinse with CO_2, co-current depressurization down to 20 kPa, and a production stage consisting of evacuation down to 10 kPa with simultaneous heating up to 70°C and a light purge step at 10 kPa. The performance of the adsorbent bed remained unaltered during the cyclic experiments. Likewise, Durán et al. [43] explored VSA cycles in a pine sawdust-based AC adsorbent bed to capture CO_2 in waste-to-energy plants. Experiments in a laboratory scale fixed-bed were complemented with simulations for a feed consisting of 8 vol% CO_2 (N_2 balance) at 50°C. At least 25 cycles were run for each VSA configuration evaluated without loss of performance of the bio-based adsorbent. The maximum CO_2 product purity reached experimentally was 39.4% (for a 96% recovery of CO_2) but simulations showed room for improvement when introducing pressure equalization steps.

Other authors, such as Shahkarami et al. [44], ran breakthrough adsorption tests at temperatures from room to 65°C and CO_2 concentrations ranging from 6 to 30 mol%, and evaluated the cyclic CO_2 adsorption performance (50 cycles) at 25°C on steam, CO_2, and KOH activated biocarbons ($\sim 5 \times 10^{-3}$ kg) from whitewood biochar in 30 mol% CO_2 in He. The highest CO_2 adsorption capacity of 1.8 mol/kg (30 mol% CO_2 and 25°C) corresponded to the KOH-activated biocarbon due to its microporous structure and high surface. The chemical and CO_2 ACs fully regenerated and preserved their performance over the 50 cycles but the adsorption capacity of the steam AC decreased after about 20 cycles, suggesting that the steam AC was not a favorable adsorbent for multi-cyclic CO_2 adsorption.

It is worth to mention that little dynamic CO_2 adsorption experimentation has been ongoing in humid flue gas conditions. Xu et al. [45] evaluated a commercial coconut shell-derived AC in a VSA cycle at 120 kPa and 60°C in humid flue gas conditions (12 vol% CO_2, 4.8 vol% H_2O, air balance). Simulations were also run to back-up the experimental results. Contrary to what would be expected, these authors observed little effect of water on the adsorption of CO_2 delivering similar recovery (~88%) and purity (~48%) of CO_2 in the presence and absence of water, although at the expense of an increase in power consumption for the humid case. However, other authors observed a small rollup in the CO_2 breakthrough curves as a consequence of H_2O adsorption when comparing the dynamic performance in dry and humid flue gas on a fresh bed of a microporous biochar AC produced from olive stones [46, 47]. The authors confirmed that the CO_2 adsorption capacity depends on the H_2O initially adsorbed on the bed but the adsorption of H_2O is little influenced by CO_2. Indeed, Durán et al. [48] observed a significant reduction in the CO_2 adsorption capacity (up to 54%) when evaluating adsorbent beds of pine sawdust-derived ACs in a fixed-bed set-up feeding ternary $CO_2/N_2/H_2O$ (8/90/2 vol%) mixtures at a high relative humidity (~60%) and highlighted the crucial role of the relative humidity of the feed in the CO_2 adsorption performance of the bed. On the other hand, the bio-adsorbents outperformed a commercial AC in conditions representative of incineration flue gas (waste-to-energy plants). Manyá et al. [49] corroborated the previous findings when evaluating the performance in adsorption-desorption cycles of CO_2 adsorbents produced from wine shoots and wheat straw pellets. They also pointed out that surface chemistry, particularly the concentration and accessibility of hydrophilic oxygen functionalities on the bio-adsorbents, could influence the water adsorption rate.

In summary, although the bio-adsorbents show lower selectivity to CO_2 over N_2 than, for instance, commercial zeolite 13X, they offer competitive advantages in terms of water tolerance and minor impact on CO_2 capture process performance under controlled humidity conditions. The possibility of tailoring the hydrophobic characteristics of the bio-adsorbent seems a promising pathway to direct the research in adsorption process application to real wet flue gases.

4.5.2 Biogas Upgrading

Biogas upgrading to produce biomethane is gaining increased interest in the current EU energy scenario facing a transition to renewable energy sources. Pressure Swing Adsorption (PSA) is a commercial technology to remove carbon dioxide from biogas but relies mainly on highly selective adsorbents to separate CO_2 and CH_4 such as zeolites or carbon molecular sieves. However, bio-based adsorbents have also been recently explored for CO_2/CH_4 separations.

When crossing the terms PSA, biogas upgrading and AC topics in a literature scientific database less than 20 publications turn up in the last five years. The number shortens when adding CO_2 capture and biomass-derived porous carbons to the topic of search [50]. Table 4.2 gathers the relevant outcome from the few papers published on biomass-derived porous carbons applied to biogas upgrading by PSA.

Despite the appealing potential demonstrated by the above-mentioned bio-based adsorbents in CO_2/CH_4 separations, validation of the adsorption process performance

TABLE 4.2
Breakthrough adsorption studies on bio-based adsorbents for biogas upgrading

Biomass precursor	V_{micro} m³/kg 10^{-3}	Scope of the study	CO_2 uptake mol/kg	Ref.
Cherry stones	0.4 ($CS\text{-}CO_2$) 0.8 ($CS\text{-}H_2O$)	Breakthrough test feeding a 50/50 vol% CO_2/CH_4 gas mixture at 30°C and 0.1–1 MPa total pressure.	At 0.5 MPa: 3.60 ($CS\text{-}CO_2$) 3.53 ($CS\text{-}H_2O$)	[51]
Moso bamboo	N.A.	Breakthrough test feeding a 40/60 vol% CO_2/CH_4 gas mixture.	2.70[a] (BC 500) 2.76[a] (BC 900)	[52]
Pine wood pellets	0.22	Breakthrough test feeding a 60/40 vol% CO_2/CH_4 gas mixture at 30°C and 0.2 MPa.	2.14	[53]
Pine sawdust pellets	0.30	Breakthrough test feeding $CH_4/CO_2/N_2/H_2O_{(v)}$ gas mixtures at 30°C and 0.1 MPa: 8.3×10^{-7} m³/s of N_2, 3.1×10^{-8} kg/s of $H_2O_{(v)}$ and 8.3×10^{-7} m³/s of CO_2/CH_4 mixtures (30/70, 50/50 and 65/35 vol%).	0.96–1.64 (depending on the CO_2/CH_4 ratio)	[54]
Silver fir sawdust	0.24[b] (HCA-200-0)	Breakthrough test feeding a 50/50 vol% CO_2/CH_4 gas mixture at 0.2 and 0.5 MPa total pressure.	6.57 at 0.5 MPa (batch equilibrium exp. in N_2)	[55]

[a] Data correspond to saturation capacity estimated from the CO_2 adsorption isotherms.
[b] Desorption pore volume from BJH.

at real biogas conditions is still pending. Only the work of Durán et al. [51] reported a comprehensive analysis of biogas upgrading on a biomass AC in both dry and humid conditions. Although a decrease in the CO_2 adsorption capacity was observed when wet biogas fed the system, the authors concluded that the presence of water vapor on the bed could promote the adsorption of CO_2 over CH_4 and lead to a more efficient separation compared to the dry case.

4.6 BIO-BASED CO_2 ADSORBENT MANUFACTURE – PROCESS DATA CASE STUDY

Biochar, an environmentally friendly product from the thermochemical processing of biomass, has gained a lot of attention recently due to its great capacity to produce various valuable products such as AC for CO_2 adsorption. Different carbonization methods are utilized to produce biochar depending on the characteristics of the biomass material used as a feedstock and the envisioned characteristics of the resulting biochar. These technologies could be summarized as: slow pyrolysis, fast pyrolysis, flash pyrolysis, intermediate pyrolysis, and vacuum pyrolysis. The residence time, heating rate, particle size, and temperature are the most influential parameters to produce high-quality biochar. Slow pyrolysis is known as the conventional carbonization method and it is the technology selected in our case study. It occurs at a slow heating rate in the absence of oxygen at a temperature range of 227–677°C, a heating rate of 0.1–1°C/s, a particle size of 5–50 mm and a residence time of 450–550 s.

Herein, the main input data of the pyrolysis unit within our case study are summarized in Table 4.3.

The activation of the biochar further enhances the porosity development and improves the adsorption capacity of the resultant AC. In the last few years, tremendous efforts have been directed to increase the capacity of the carbons to adsorb CO_2 selectively. The process essentially responds to the micropore-volume filling by physical adsorption, although the CO_2 uptake can improve by chemisorption on surface sites. In this context, the current strategy of bio-based CO_2 adsorbent development focuses primarily on advanced materials with a tailored porous structure and/or functionalized surface. Our case study puts the focus on physical activation using CO_2 as the activating agent, opening the possibility of an alternative use for the captured CO_2. The relevant attributes for CO_2 adsorption of the AC produced from forest biomass in our case study are summarized in Table 4.4.

TABLE 4.3
Input parameters in the modeling of the case study pyrolysis unit [52]

Molecular formula for the biomass feedstock ("solid")	$CH_{1.46}O_{0.689}$
Molecular formula for the biochar	$CH_{0.153}O_{0.0324}$
Molecular weight of the biochar	12.70 kg/kmol
Biochar yield	25.6 wt% [db]
Yield of carbon dioxide purification (biochar basis)	0.1 kg CO_2/kg biochar
Yield of carbon dioxide purification (solid basis)	0.63 kg CO_2/kmol solid [db]
Temperature of pyrolysis	800°C
Heat for biomass treatment	40.1 MJ/kmol solid [db]
Heat for solid treatment (isolated carbon dioxide basis)	63.65 MJ/kg CO_2
Products composition pyrolysis [1]	
Processing gas (on a 100 g biomass db):	
- CO_2	13.6 wt%
- CO	6.7 wt%
- H_2	0.3 wt%
- CH_4	1.6 wt%
- C_2H_4	0.2 wt%
- H_2O	24.6 wt%
Wastes:	
- Liquid product (tar + water)	26.2 wt%

[db] dry basis.

TABLE 4.4
Main characteristics of the activation for the production of bio-adsorbent

Activating agent	$CO_{2(g)}$
Activation temperature	800°C
Activation yield	25
Narrow micropore volume (<0.7 nm)	0.27 dm^3/kg adsorbent
Average narrow micropore size	0.56 nm
Maximum CO_2 adsorption capacity (100 kPa, 25°C)	2.1 mol CO_2/kg adsorbent

It has been demonstrated that CO_2 adsorption on ACs at atmospheric pressure is effective only when the average pore width is smaller than the molecular size of the adsorbate. High pore volumes and particularly narrow micropores increase the CO_2 uptake. Therefore, the design of bio-based adsorbents with a large volume of narrow micropores is essential to achieve high CO_2 uptake at ambient conditions.

REFERENCES

[1] Basu, P. *Biomass Gasification and Pyrolysis*, Academic Press, 2010, ISBN 9780123749888, https://doi.org/10.1016/C2009-0-20099-7

[2] Onay, O.; Kockar, O. M. Slow, fast and flash pyrolysis of rapeseed. *Renewable Energy* **2003**, *28*, 2417–2433, https://doi.org/10.1016/S0960–1481(03)00137-X.

[3] Jain, A.; Balasubramanian, R.; Srinivasan, M. P. Hydrothermal conversion of biomass waste to activated carbon with high porosity: A review. *Chemical Engineering Journal* **2016**, *283*, 789–805, https://doi.org/10.1016/j.cej.2015.08.014.

[4] Román, S.; Valente Nabais, J. M.; Ledesma, B.; González, J. F.; Laginhas, C.; Titirici, M. M. Production of low-cost adsorbents with tunable surface chemistry by conjunction of hydrothermal carbonization and activation processes. *Microporous and Mesoporous Materials* **2013**, *165*, 127–133, https://doi.org/10.1016/j.micromeso.2012.08.006.

[5] Reza, M. T.; Lynam, J. G.; Vasquez, V. R.; Coronella, C. J. Pelletization of biochar from hydrothermally carbonized wood. *Environmental Progress & Sustainable Energy* **2012**, *31*, 225–234, https://doi.org/10.1002/ep.11615.

[6] Álvarez Merino, M. A.; Carrasco Marín, F.; Maldonado Hódar, F. J. *Desarrollo y aplicaciones de materiales avanzados de carbón*; Universidad Internacional de Andalucía, 2014.

[7] Samanta, A.; Zhao, A.; Shimizu, G. K. H.; Sarkar, P.; Gupta, R. Post-combustion CO_2 capture using solid sorbents: A review. *Industrial and Engineering Chemistry Research* **2012**, *51*, 1438–1463, https://doi.org/10.1021/ie200686q.

[8] Bui, M.; Adjiman, C. S.; Bardow, A.; Anthony, E. J.; Boston, A.; Brown, S.; Fennell, P. S.; Fuss, S.; Galindo, A.; Hackett, L. A.; Hallett, J. P.; Herzog, H. J.; Jackson, G.; Kemper, J.; Krevor, S.; Maitland, G. C.; Matuszewski, M.; Metcalfe, I. S.; Petit, C.; Puxty, G.; Reimer, J.; Reiner, D. M.; Rubin, E. S.; Scott, S. A.; Shah, N.; Smit, B.; Trusler, J. P. M.; Webley, P.; Wilcox, J.; Mac Dowell, N. Carbon capture and storage (CCS): The way forward. *Energy & Environmental Science* **2018**, *11*, 1062–1176, https://doi.org/10.1039/C7EE02342A.

[9] Lehmann, J.; Stephen, J. *Biochar for Environmental Management: Science, Technology and Implementation*; Routledge, 2015; Vol. 1.

[10] Rouzitalab, Z.; Maklavany, D. M.; Jafarinejad, S.; Rashidi, A. Lignocellulose-based adsorbents: A spotlight review of the effective parameters on carbon dioxide capture

process. *Chemosphere* **2020**, *246*, 125756, https://doi.org/10.1016/j.chemosphere.2019.125756.

[11] Wang, J.; Huang, L.; Yang, R.; Zhang, Z.; Wu, J.; Gao, Y.; Wang, Q.; O'Hare, D.; Zhong, Z. Recent advances in solid sorbents for CO_2 capture and new development trends. *Energy & Environmental Science* **2014**, *7*, 3478–3518, https://doi.org/10.1039/C4EE01647E.

[12] Querejeta, N.; Gil, M. V.; Pevida, C.; Centeno, T. A. Standing out the key role of ultramicroporosity to tailor biomass-derived carbons for CO_2 capture. *Journal of CO_2 Utilization* **2018**, *26*, 1–7, https://doi.org/10.1016/j.jcou.2018.04.016.

[13] Vargas, D. P.; Giraldo, L.; Silvestre-Albero, J.; Moreno-Piraján, J. C. CO_2 adsorption on binderless activated carbon monoliths. *Adsorption* **2011**, *17*, 497–504, https://doi.org/10.1007/s10450-010-9309-z.

[14] Plaza, M. G.; Pevida, C.; Martín, C. F.; Fermoso, J.; Pis, J. J.; Rubiera, F. Developing almond shell-derived activated carbons as CO_2 adsorbents. *Separation and Purification Technology* **2010**, *71*, 102–106, https://doi.org/10.1016/j.seppur.2009.11.008.

[15] González, A. S.; Plaza, M. G.; Rubiera, F.; Pevida, C. Sustainable biomass-based carbon adsorbents for post-combustion CO_2 capture. *Chemical Engineering Journal* **2013**, *230*, 456–465, https://doi.org/10.1016/j.cej.2013.06.118.

[16] Yang, R.; Liu, G.; Li, M.; Zhang, J.; Hao, X. Preparation and N_2, CO_2 and H_2 adsorption of super activated carbon derived from biomass source hemp (Cannabis sativa L.) stem. *Microporous and Mesoporous Materials* **2012**, *158*, 108–116, https://doi.org/10.1016/j.micromeso.2012.03.004.

[17] Álvarez-Gutiérrez, N.; Gil, M. V.; Rubiera, F.; Pevida, C. Cherry-stones-based activated carbons as potential adsorbents for CO_2/CH_4 separation: Effect of the activation parameters. *Greenhouse Gases: Science and Technology* **2015**, *5*, 812–825, https://doi.org/10.1002/ghg.1534.

[18] Yang, H.; Gong, M.; Chen, Y. Preparation of activated carbons and their adsorption properties for greenhouse gases: CH_4 and CO_2. *Journal of Natural Gas Chemistry* **2011**, *20*, 460–464, https://doi.org/10.1016/S1003–9953(10)60232-0.

[19] Ello, A. S.; De Souza, L. K. C.; Trokourey, A.; Jaroniec, M. Coconut shell-based microporous carbons for CO_2 capture. *Microporous and Mesoporous Materials* **2013**, *180*, 280–283, https://doi.org/10.1016/j.micromeso.2013.07.008.

[20] Abdeljaoued, A.; Querejeta, N.; Durán, I.; Álvarez-Gutiérrez, N.; Pevida, C.; Chahbani, M. H. Preparation and evaluation of a coconut shell-based activated carbon for CO_2/CH_4 separation. *Energies* **2018**, *11*, https://doi.org/10.3390/en11071748.

[21] Plaza, M. G.; González, A. S.; Pevida, C.; Pis, J. J.; Rubiera, F. Valorisation of spent coffee grounds as CO_2 adsorbents for postcombustion capture applications. *Applied Energy* **2012**, *99*, 272–279, https://doi.org/10.1016/j.apenergy.2012.05.028.

[22] Plaza, M. G.; González, A. S.; Pevida, C.; Rubiera, F. Green coffee based CO_2 adsorbent with high performance in postcombustion conditions. *Fuel* **2015**, *140*, 633–648, https://doi.org/10.1016/j.fuel.2014.10.014.

[23] Querejeta, N.; Gil, M. V.; Rubiera, F.; Pevida, C. Sustainable coffee-based CO_2 adsorbents: Toward a greener production via hydrothermal carbonization. *Greenhouse Gases: Science and Technology* **2018**, *8*, 309–323, https://doi.org/10.1002/ghg.1740.

[24] Plaza, M. G.; Pevida, C.; Arias, B.; Fermoso, J.; Casal, M. D.; Martín, C. F.; Rubiera, F.; Pis, J. J. Development of low-cost biomass-based adsorbents for postcombustion CO_2 capture. *Fuel* **2009**, *88*, 2442–2447, https://doi.org/10.1016/j.fuel.2009.02.025.

[25] Puig-Gamero, M.; Esteban-Arranz, A.; Sanchez-Silva, L.; Sánchez, P. Obtaining activated biochar from olive stone using a bench scale high-pressure thermobalance. *Journal of Environmental Chemical Engineering* **2021**, *9*, 105374, https://doi.org/10.1016/j.jece.2021.105374.

[26] Sher, F.; Iqbal, S. Z.; Albazzaz, S.; Ali, U.; Mortari, D. A.; Rashid, T. Development of biomass derived highly porous fast adsorbents for post-combustion CO_2 capture. *Fuel* **2020**, *282*, 118506, https://doi.org/10.1016/j.fuel.2020.118506.

[27] Deng, S.; Wei, H.; Chen, T.; Wang, B.; Huang, J.; Yu, G. Superior CO_2 adsorption on pine nut shell-derived activated carbons and the effective micropores at different temperatures. *Chemical Engineering Journal* **2014**, *253*, 46–54, https://doi.org/10.1016/j.cej.2014.04.115.

[28] Li, D.; Ma, T.; Zhang, R.; Tian, Y.; Qiao, Y. Preparation of porous carbons with high low-pressure CO_2 uptake by KOH activation of rice husk char. *Fuel* **2015**, *139*, 68–70, https://doi.org/10.1016/j.fuel.2014.08.027.

[29] Creamer, A. E.; Gao, B.; Zhang, M. Carbon dioxide capture using biochar produced from sugarcane bagasse and hickory wood. *Chemical Engineering Journal* **2014**, *249*, 174–179, https://doi.org/10.1016/j.cej.2014.03.105.

[30] Chomiak, K.; Gryglewicz, S.; Kierzek, K.; Machnikowski, J. Optimizing the properties of granular walnut-shell based KOH activated carbons for carbon dioxide adsorption. *Journal of CO_2 Utilization* **2017**, *21*, 436–443, https://doi.org/10.1016/j.jcou.2017.07.026.

[31] Asadi-Sangachini, Z.; Galangash, M. M.; Younesi, H.; Nowrouzi, M. The feasibility of cost-effective manufacturing activated carbon derived from walnut shells for large-scale CO_2 capture. *Environmental Science and Pollution Research* **2019**, *26*, 26542–26552, https://doi.org/10.1007/s11356-019-05842-3.

[32] Wang, Y.; Liu, C.; Zhou, Y. Preparation and adsorption performances of mesopore-enriched bamboo activated carbon. *Frontiers of Chemical Engineering in China* **2008**, *2*, 473–477, https://doi.org/10.1007/s11705-008-0081–5.

[33] Wei, H.; Deng, S.; Hu, B.; Chen, Z.; Wang, B.; Huang, J.; Yu, G. Granular bamboo-derived activated carbon for high CO_2 adsorption: The dominant role of narrow micropores. *ChemSusChem* **2012**, *5*, 2354–2360, https://doi.org/10.1002/cssc.201200570.

[34] Khuong, D. A.; Nguyen, H. N.; Tsubota, T.; Khuongab, D. A.; Nguyen, H. N.; Tsubota, T.; Khuong, D. A.; Nguyen, H. N.; Tsubota, T. Activated carbon produced from bamboo and solid residue by CO_2 activation utilized as CO_2 adsorbents. *Biomass and Bioenergy* **2021**, *148*, 106039, https://doi.org/10.1016/j.biombioe.2021.106039.

[35] Heidari, A.; Younesi, H.; Rashidi, A.; Ghoreyshi, A. A. Adsorptive removal of CO_2, on highly microporous activated carbons prepared from Eucalyptus camaldulensis wood: Effect of chemical activation. *Journal of the Taiwan Institute of Chemical Engineers* **2014**, *45*, 579–588, https://doi.org/10.1016/j.jtice.2013.06.007.

[36] Huang, Y. F.; Chiueh, P. Te; Lo, S. L. CO_2 adsorption on biochar from co-torrefaction of sewage sludge and leucaena wood using microwave heating. *Energy Procedia* **2019**, *58*, 4435–4440, https://doi.org/10.1016/j.egypro.2019.01.772.

[37] Li, Y.; Ruan, G.; Jalilov, A. S.; Tarkunde, Y. R.; Fei, H.; Tour, J. M. Biochar as a renewable source for high-performance CO_2 sorbent. *Carbon* **2016**, *107*, 344–351, https://doi.org/10.1016/j.carbon.2016.06.010.

[38] Zhu, X.-L. L.; Wang, P.-Y. Y.; Peng, C.; Yang, J.; Yan, X.-B. Bin Activated carbon produced from paulownia sawdust for high-performance CO_2 sorbents. *Chinese Chemical Letters* **2014**, *25*, 929–932, https://doi.org/10.1016/j.cclet.2014.03.039.

[39] Plaza, M. G.; Durán, I.; Rubiera, F.; Pevida, C. CO_2 adsorbent pellets produced from pine sawdust: Effect of coal tar pitch addition. *Applied Energy* **2015**, *144*, 182–192, https://doi.org/10.1016/j.apenergy.2014.12.090.

[40] Tian, Z.; Qiu, Y.; Zhou, J.; Zhao, X.; Cai, J. The direct carbonization of algae biomass to hierarchical porous carbons and CO_2 adsorption properties. *Materials Letters* **2016**, *180*, 162–165, https://doi.org/10.1016/j.matlet.2016.05.169.

[41] Sevilla, M.; Falco, C.; Titirici, M.-M.; Fuertes, A. B. High-performance CO_2 sorbents from algae. *RSC Advances* **2012**, *2*, 12792, https://doi.org/10.1039/c2ra22552b.

[42] Durán, I.; Rubiera, F.; Pevida, C. Microalgae: Potential precursors of CO_2 adsorbents. *Journal of CO_2 Utilization* **2018**, *26*, 454–464, https://doi.org/10.1016/j.jcou.2018.06.001.
[43] Durán, I.; Rubiera, F.; Pevida, C. Vacuum swing CO_2 adsorption cycles in waste-to-energy plants. *Chemical Engineering Journal* **2020**, *382*, 122841, https://doi.org/10.1016/j.cej.2019.122841.
[44] Shahkarami, S.; Azargohar, R.; Dalai, A. K.; Soltan, J. Breakthrough CO_2 adsorption in bio-based activated carbons. *Journal of Environmental Sciences (China)* **2015**, *34*, 68–76, https://doi.org/10.1016/j.jes.2015.03.008.
[45] Xu, D.; Xiao, P.; Zhang, J.; Li, G.; Xiao, G.; Webley, P. A.; Zhai, Y. Effects of water vapour on CO_2 capture with vacuum swing adsorption using activated carbon. *Chemical Engineering Journal* **2013**, *230*, 64–72, https://doi.org/10.1016/j.cej.2013.06.080.
[46] Plaza, M. G.; Durán, I.; Querejeta, N.; Rubiera, F.; Pevida, C. Experimental and simulation study of adsorption in postcombustion conditions using a microporous biochar. 1. CO_2 and N_2 adsorption. *Industrial and Engineering Chemistry Research* **2016**, *55*, 3097–3112, https://doi.org/10.1021/acs.iecr.5b04856.
[47] Plaza, M. G.; Durán, I.; Querejeta, N.; Rubiera, F.; Pevida, C. Experimental and simulation study of adsorption in postcombustion conditions using a microporous biochar. 2. H_2O, CO_2, and N_2 adsorption. *Industrial & Engineering Chemistry Research* **2016**, *55*, 6854–6865, https://doi.org/10.1021/acs.iecr.6b01720.
[48] Durán, I.; Rubiera, F.; Pevida, C. Separation of CO_2 in a solid waste management incineration facility using activated carbon derived from pine sawdust. *Energies* **2017**, *10*, 827, https://doi.org/10.3390/en10060827.
[49] Manyà, J. J.; García-Morcate, D.; González, B. Adsorption performance of physically activated biochars for postcombustion CO_2 capture from dry and humid flue gas. *Applied Sciences (Switzerland)* **2020**, *10*, 376, https://doi.org/10.3390/app10010376.
[50] Bernardo, M.; Lapa, N.; Fonseca, I.; Esteves, I. A. A. C. Biomass valorization to produce porous carbons: Applications in CO_2 capture and biogas upgrading to biomethane—A mini-review. *Frontiers in Energy Research* **2021**, *9*, 74, https://doi.org/10.3389/fenrg.2021.625188.
[51] Durán, I.; Álvarez-Gutiérrez, N.; Rubiera, F.; Pevida, C. Biogas purification by means of adsorption on pine sawdust-based activated carbon: Impact of water vapor. *Chemical Engineering Journal* **2018**, *353*, 197–207, https://doi.org/10.1016/j.cej.2018.07.100.
[52] Kodera Y.; Kaiho M. Model calculation of heat balance of wood pyrolysis. *Journal of the Japan Institute of Energy* **2016**, *95*, 881–889, https://doi.org/10.3775/jie.95.881.

5 CO_2 hydrogenation into dimethyl ether *Conventional and innovative catalytic routes*

Giuseppe Bonura, Serena Todaro, Catia Cannilla, and Francesco Frusteri

CONTENTS

5.1 INTRODUCTION

5.1.1 DME AS AN ALTERNATIVE CLEAN FUEL

Dimethyl ether (DME), also known as methoxymethane, wood ether, dimethyl oxide or methyl ether, is a very useful molecule. DME is the simplest ether currently used as a pressurizing agent in spray cans, paints and insecticides instead of the previously applied chlorofluorocarbons. DME is also an important chemical intermediate for the production of widely used chemicals, such as methyl acetate, diethyl sulfate, gasoline and light olefins. It can significantly reduce the carbon footprint of the transportation sector and beyond (1) as an energy-dense, cost-effective

DOI: 10.1201/9781003336587-5

carrier to move renewable hydrogen, (2) as a blending agent for propane and (3) as a diesel replacement [1–7]. High cetane number (55–60) and high-efficiency compression ignition fuel are the main features that make DME a promising alternative diesel fuel. When it is pressurized above 0.5 MPa, it condenses to the liquid phase. From this point of view, it is an environmentally benign fuel neither toxic nor carcinogenic. During combustion, soot impurities are not released, unlike diesel oil. DME's easy handling properties make fueling and infrastructure relatively simple and inexpensive. Diesel engines require only minor modifications to be able to burn DME, saving ~30% greater efficiency of compression engines fed by fossil diesel, also accounting for low emissions of NO_x, SO_x and particulate in the atmosphere. DME can be synthesized from various feedstock, starting from natural gas, coal and biomass.

As displayed in Figure 5.1, it can be synthesized from syngas or carbon dioxide using either two-step or one-step processes [8–12]. In the first step, methanol is directly formed from the conversion of CO or CO_2 using copper-based catalysts; in the second step, DME production requires acid functionalities for the subsequent dehydration of methanol. In the one-step process, the catalyst functionalities related to methanol synthesis and methanol dehydration are combined in a hybrid system within a single reactor, so to alleviate the thermodynamic constraints of the hydrogenation step and leading to higher CO_2 conversion rates.

Despite this, hydrogen production from renewables remains an open challenge, and fossil fuels are projected to remain the mainstay of electricity generation in many countries, especially where they are the principal indigenous and economically viable energy sources.

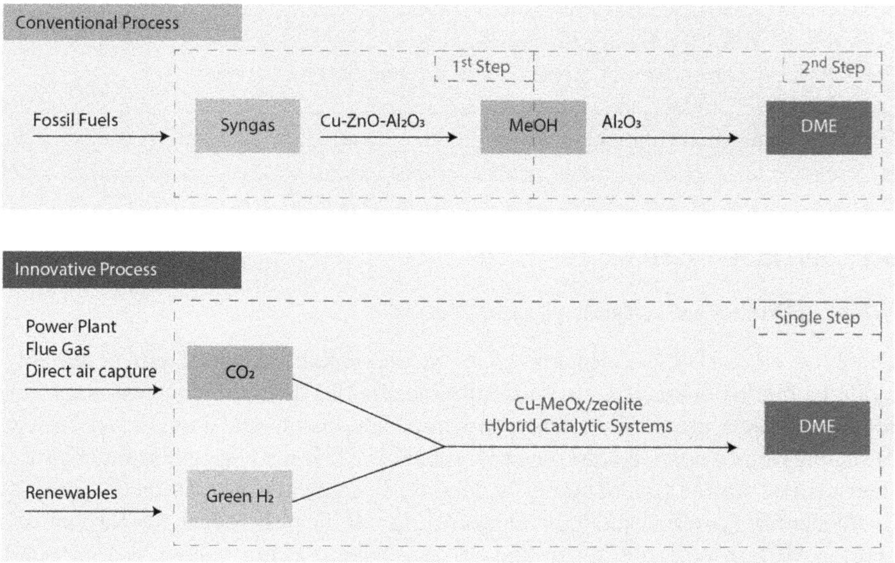

FIGURE 5.1 Scheme of conventional and innovative process routes to DME.

5.1.2 CO_2 UTILIZATION: CONSIDERATIONS ON CHEMICAL ACTIVATION

Since CO_2 represents the last stage of carbon oxidation, it is problematic to be activated because of its thermodynamic and kinetic stability. As a result of these, an external energy effort and a suitable catalyst are required for the conversion of CO_2 into value-added products. Aided by external energy sources such as temperature, light or electricity, a unique environment can be created to activate this rather inert molecule. If the CO_2 used in the production of transport fuels is sourced from industrial processes, the overall result is a reduction in the consumption of fossil resources. In this case, CO_2 serves to broaden the raw materials base of the chemical industry and CO_2 emissions are reduced as conventional fossil-based fuels do not need to be used.

From a thermodynamic point of view, Figure 5.2 displays a decreasing trend of CO_2 conversion (X_{CO_2}) when temperature increases, with a maximum conversion value above 60% at 180°C and 5 MPa, while resulting as low as 17% at around 240°C and 1 MPa. Within the investigated temperature range, the DME equilibrium selectivity gradually decreases, with a more significant drop at a reaction pressure as low as 1 MPa, being counterbalanced by a dramatic increase of CO which is less marked at higher pressure. In general, the effect of temperature appears less pronounced at any pressure on the methanol production. On the whole, such findings imply that CO_2 hydrogenation reaction should be carried out at low temperature and high pressure in order to maximize CO_2 conversion and selectivity to DME.

Obviously, potential disparities between thermodynamic predictions and kinetics are to be taken into account, depending on the feed compositions (in terms of H_2/CO_2 ratio and possible co-feeding of CO and/or water) as well as on specific catalyst.

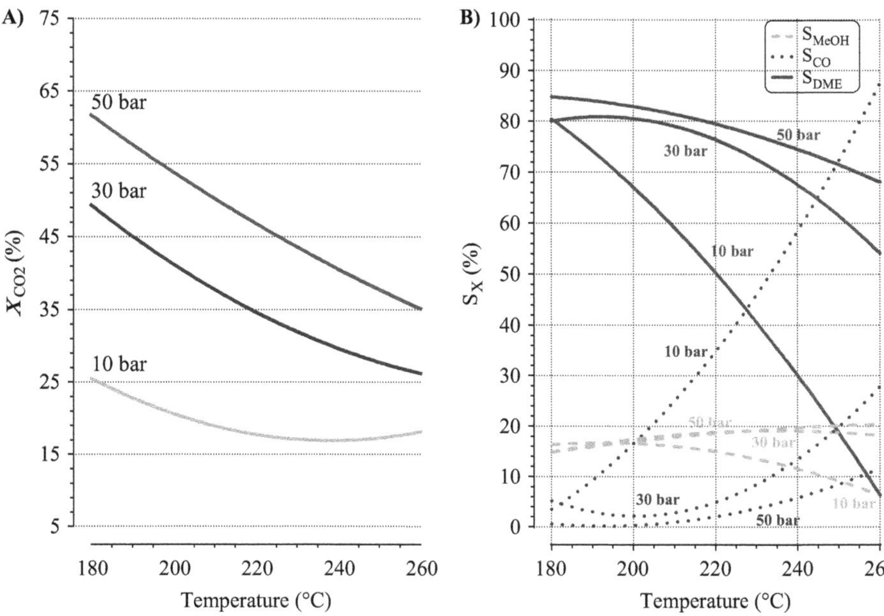

FIGURE 5.2 Equilibrium values expected on the basis of thermodynamic analysis of direct DME synthesis from CO_2 hydrogenation (CO_2/H_2, 1/3 mol/mol).

The hydrogenation of CO_2 into DME has been recognized as one of the most economical ways to recycle it [13–15]. The process involves the following reactions:

Methanol formation

$$CO_2 + 3H_2 \rightarrow CH_3OH + H_2O \tag{5.1}$$

Methanol dehydration

$$2CH_3OH \rightarrow CH_3OCH_3 + H_2O \tag{5.2}$$

Reverse water-gas shift (rWGS)

$$CO_2 + H_2 \rightarrow CO + H_2O \tag{5.3}$$

Net reaction

$$2CO_2 + 6H_2 \rightarrow CH_3OCH_3 + 3H_2O \tag{5.4}$$

A disadvantage of this process relies on using a large amount of hydrogen. It is not economically convenient because of its cost, but the economical production of hydrogen by renewable resources could radically alter this scenario. Currently, most hydrogen (about 95%) is produced by fossil source-based methods, among which steam reforming of methane, partial oxidation, auto-thermal and dry-reforming [13, 14]. Biomass is considered an excellent alternative to petroleum. Especially, biomass gasification may be adopted as an appropriate technology to produce syngas and then hydrogen without combustion [16]. This process is conducted at high temperatures in the presence of oxygen, air and steam. Thanks to adsorbent systems or membranes, it is possible to separate hydrogen from the gas stream. Other technologies have been evaluated for hydrogen production from biomass, such as biomass fermentation, using the anaerobic bacteria or green algae, pyrolysis and supercritical water gasification, although the applications are inadequate to large scale. Hydrogen gas with high purity can be also produced from water by various technologies [16]. Water electrolysis is the most commonly used in which water is converted into hydrogen and oxygen at low temperatures using electrical energy. The price of the process strongly depends on the cost of electricity. The best way is the use of electrical energy derivable from renewable sources such as wind or sun hydro-power or biomass-derived sources [13].

Despite this, fossil fuels are projected to remain the mainstay of electricity generation in many countries, especially where fossil fuels are the principal indigenous and economically viable sources of energy. Hydrogen can be produced from various renewable sources, such as solar, wind, hydro-power or biomass-derived sources.

5.2 CONVENTIONAL ROUTES FOR DME PRODUCTION

DME is typically produced from syngas conversion in a two-step process. The two-step process is considered the most mature route from an industrial perspective and is currently employed by companies such as Haldor Topsoe, Toyo Engineering, Oberon Fuels, BioDME and Lurgi Air Products, in which syngas represents the feedstock [17]. The process involves first methanol formation in a reactor and the subsequent dehydration of methanol in another reactor. This separate dehydration step is reported to require low capital investment, provided there is high feedstock availability. Syngas is a feedstock produced either by gasification of coal, biomass or by steam reforming of natural gas. The syngas obtained by steam reforming becomes more suitable since it does not contain impurities such as sulfur species or heavy metals. In the first step, syngas is converted into methanol over a methanol catalyst generally based on copper. Methanol is purified and then converted to DME in another reactor over acid catalysts (zeolites, alumina or phosphoric acid modified γ-Al_2O_3).

Since methanol synthesis, water/gas shift and methanol dehydration are exothermic processes, then DME production is thermodynamically favorable at lower temperatures. The reaction at higher temperatures can also favor the formation of other hydrocarbons or coke and the sintering of metal crystals, leading to the deactivation of the catalyst.

The following approach leads to satisfactory selectivity under mild operating conditions, high purity of the products, which do not justify the high investment costs. The formation of methanol and its subsequent dehydration requires two different catalysts and reactors. Moreover, due to the thermodynamic limitations of methanol formation, low conversion levels are obtained.

5.2.1 Methanol Phase Catalyst

In 1923, the first commercial catalyst (i.e., ZnO-Cr_2O_3) for the production of methanol starting from syngas was proposed by BASF, active at high pressure and temperature [18]. The main problem of this catalyst was the sensitivity to poisoning due to impurities derived from syngas. In 1940, another methanol catalyst from syngas was proposed based on the use of Cu as an active phase and ZnO as a promoter to increase the catalyst activity. Nevertheless, this catalyst is less active in the synthesis of methanol from CO_2. The CO_2 conversion does not exceed 20% due to the difficulty of activating CO_2 molecule. In 1960, an efficient low-temperature CuO/ZnO/-Al_2O_3 catalyst was proposed by Imperial Chemistry Industry (ICI) in which ZnO was used as a support thanks to high stability to increase the dispersion of Cu and to stabilize the active phase with an appropriate ratio Cu^+/Cu^0 suitable to generate oxygen defects at the interface [19]. This is still today the most widely used industrial catalyst for methanol production. The incorporation of noble metals to the classical Cu/ZnO/Al_2O_3 catalysts for CO_2 hydrogenation to methanol was widely studied. In particular, the addition of noble metals such as Pd or Au enhances the system activity but not justifies the high price associated with the preparation of the catalyst. Other studies about the correlation between the oxide support with catalytic properties of the system are evaluated. Typically, a suitable support material is based on modulating surface acidity/basicity to affect the activation of CO_2 and H_2 and adjustable textural properties to promote mass transfer. The substitution of Al_2O_3 by various metal oxides such as ZrO_2, CeO_2, SiO_2 or TiO_2 has been widely investigated [11, 12]. Among these metal oxides, ZrO_2 has proven the best catalytic behavior. ZrO_2

enhances the basicity of the system, which favors the CO_2 adsorption increasing the selectivity to methanol. Furthermore, the high stability upon reduction, the significant surface area and pore volume make it the best candidate in terms of textural properties and durability [10].

Recently, another element has attracted much attention; in particular, In_2O_3 displays promising high methanol selectivity. The key intermediates (*HCOO) necessary for the synthesis of methanol are more stable on an In_2O_3-based surface containing defects compared to a Cu-based surface. This potential allows to suppress the formation of CO and therefore to favor the selectivity to methanol [20, 21]. The indium-based catalysts that have been tested consist of only the active phase (In_2O_3), the active phase on a zirconia support (In_2O_3/ZrO_2) and also doped with yttrium or lanthanum. The In_2O_3/ZrO_2 catalyst showed higher selectivity to methanol than Cu- and Pd-based catalysts, thanks to a higher adsorption energy of CO_2 on the zirconia surface, the oxygen vacancies created on In_2O_3 and a considerable concentration of oxygen atoms near the vacancies. Sun et al. [22] tested the In_2O_3 catalyst and observed that it has good thermal and structural stability at temperatures below 500°C. They achieved the best methanol yields and formation rates at 330°C and 4 MPa, which decrease beyond 350°C as the rWGS reaction is favored. They highlight that no pure oxide except In_2O_3 shows a high activity for the synthesis of methanol from the hydrogenation of CO_2; in fact, it is the only oxide that is able to inhibit the rWGS reaction, for this reason, it is the only one that allows to obtain a high selectivity (39.7%) even at temperatures above 300°C [20, 23].

5.2.2 Methanol Dehydration Catalyst

Methanol dehydration is an acid-catalyzed reaction. The selection of acidic support deserves special attention; it should have some particular highlights, like high surface area, low cost and good mechanical and thermal stability. Moreover, the acidic phase also causes coke formation and it gets worse when stronger acidic systems are used; therefore it becomes necessary to modulate the surface acidity [24].

The most investigated solid acid phase is γ-Al_2O_3 because it possesses properties as those just described. Unfortunately, in contrast to CO hydrogenation, a massive amount of water during CO_2 hydrogenation is formed, especially in the direct process and γ-Al_2O_3 due to its hydrophilic surface tends to strongly adsorb water, causing a rapid loss in its catalytic activity for the blocking of the active centers of the catalyst [25].

Heteropolyacids (HPAs) could be an intriguing option in contrast to utilizing γ-Al_2O_3. They can be described using the formula $H_{8-n}[Xn^{+}M_{12}O_{40}]$, in which "X" is the central atom (e.g., Si^{4+}, Al^{3+}, etc.) and "M" is the metal ion. Because the high Brønsted acidity displayed from these materials, HPAs offer catalytic performances usually better than other solid catalysts especially at low temperature, as attributed to a higher acid site strength [26, 27].

Furthermore, zeolites constitute another class of catalysts of acidic character. The strength and the type of the acid centers on the catalyst surface are related to the structure of zeolite and its Si/Al ratio [28–30]. Commonly, H-ZSM-5 is used because of its better hydrophobic character than Al_2O_3. However, due to the high reaction temperature (above 270°C), hydrocarbons can be observed among the formed products. To improve selectivity to DME prompted by a high methanol conversion rate, it is needed to decrease the amount of strong acid centers maintaining the total acidity at high level.

5.3 DIRECT SYNTHESIS OF DME

In the last years, a one-step process has been proposed as a promising alternative to the two-step process, aiming at addressing syngas conversion into methanol and methanol dehydration to DME in the same reactor over a bifunctional/hybrid catalyst, where the functionalities of methanol synthesis and methanol dehydration are properly integrated [8, 9, 13–15]. Indeed, the one-step process overcomes the thermodynamic constraints of methanol production due to the continuous methanol stripping in the following dehydration step on a solid acid system, thus leading to higher equilibrium conversions than syngas conversion to methanol. Moreover, since direct synthesis occurs in a single reactor, lower production costs in terms of operational costs and reduced investment are required [14]. Recently, studies explore the feasibility of producing DME by direct synthesis starting from CO_2 hydrogenation to mitigate carbon dioxide emissions. The direct synthesis of DME starting from H_2/CO_2 gas mixture demands new technologies and catalysts to reduce the energy demand of the process and a suitable extent of conversion [31–35]. In the direct synthesis, catalyst plays the most crucial role in the process. As represented in Figure 5.3, the current challenge is to design a catalyst system in which a close interaction among metal-oxide sites and acid sites is realized.

The methods for catalyst preparation, such as impregnation, physical mixing or coprecipitation, have an essential effect on the process performance. In this respect, a main distinction can be operated between mechanical mixtures and hybrid catalysts,

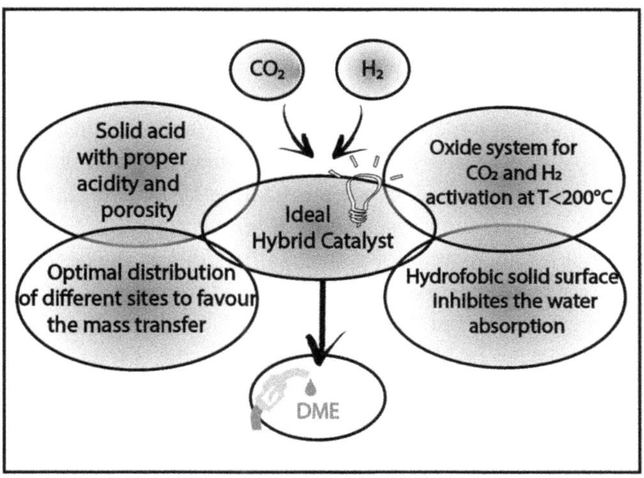

FIGURE 5.3 Concept of an ideal hybrid catalyst system for the direct synthesis of DME from CO_2/H_2 mixtures.

by considering the simple physical mixing between two preformed phases as belonging to the first class, while the chemical generation of sites of different nature being associated to an effective hybridization of the catalytic system. If on one hand, metal-oxide sites result to be specific in the activation of CO_2 and H_2, the acid functionality allows for the dehydration of the intermediate methanol, adsorbed on the catalyst surface [13–15]. The acid sites are generally classified as Brønsted and Lewis sites, whose concentration and strength greatly influence catalytic activity. As a rule, the formation of DME positively benefits from a high concentration of medium-weak acid sites, while the presence of strong acid sites generally promotes the formation of olefins [30].

Conventionally, the bifunctional catalysts are prepared by mechanical/physical mixing of preformed dry powders of a methanol phase and an acid carrier. The mechanical combination has some limitations, mainly related to: not homogeneous distribution of active sites, catalytic phases spatially separated, not fully repeatability of the mixing procedure and mass transfer restraints. A combination of the two types of catalytic sites can be implemented by means of a single pot synthesis through which the active phases are generated during the preparation with a close interaction with each other improving the activation of both CO_2 and hydrogen. These systems are typically prepared by co-precipitation or impregnation of the metal precursors in the presence or over the solid acid phase [9–11].

5.3.1 Different design of DME reactors

In recent times, many researchers have investigated how to minimize energy consumption of chemical processes through a thorough reactor design. The main challenge is to know and predict reactor behaviors, comprising temperature profiles, under different situations for scale-up.

5.3.1.1 Fixed-beds reactors

The most common reactors for catalytic conversion of syngas as well as hydrogenation of CO_2 are the fixed-bed ones, due to their easy management and lower costs either at laboratory or larger scale [36]. A fixed bed reactor consists of a cylindrical tube that contains catalyst pellets with reactants flowing through. The flow interacts with the catalytic system across the length of the tube, converting into products. In these systems, an important issue is the optimization of the heat transfer through the reactor. When the catalyst is exposed to high temperatures, hot spots are formed, leading to irreversible deactivation by sintering. In addition, the occurrence of excessive temperatures affects conversion, selectivity and lifecycle of catalysts. For catalytic processes in which the heat of reaction is small, adiabatic fixed-bed reactors can be used where an external insulating jacket surrounds the reactor.

5.3.1.2 Slurry reactors

In 1991, Air Product and Chemicals, Inc. proposed the direct synthesis of DME in a slurry reactor a kind of reactor where fine catalyst particles are suspended in a solvent, through which a gas is bubbled operating in either semi-batch or continuous mode [37]. In this case, the temperature control is aided by the huge heat capacity of the solvent that is cooled by heat exchange tubes in the reactor. For the synthesis reaction to occur, the reactants should be moved from gas bubbles to liquid phase solvent and then to catalyst particles. However, an essential difficulty of this type of reactor is the low mass transfer intensity of reagents toward the active centers of the catalyst. Low solubility and diffusivity in the liquid phase, consequently decrease the overall reaction rate. The equipment required a recycling system and a gas–liquid separator. Furthermore, the loss of catalyst particles formed by attrition in the reactor is another problem that limits the usage in DME production [38].

5.3.1.3 Fluidized-bed reactors

In a fluidized-bed reactor, an upward-flowing stream of fluid passed through the catalyst at speeds to suspend the solid particles; this process is called fluidization. Xiao and Lu [39] proposed an ideal fluidized-bed reactor for DME synthesis. In this system, the gas–solid mass-transfer resistance is so small that it can be ignored on the contrary in fixed-bed and slurry reactors. Thanks to the rapid mixing of catalyst particles, excellent temperature control is achieved. Some problems are the difficulty in operability, difficult separation of the catalyst particles from the exhaust gas and erosion problematic due to the high linear speeds leading to high capital costs. DME synthesis in fluidized bed reactors is yet in the stage of laboratory testing; its feasibility has not been completely demonstrated.

5.4 PROCESS PARAMETERS OF DIRECT CO_2 HYDROGENATION – CASE STUDY FOR DME PRODUCTION

As a case study, we investigated the direct hydrogenation of CO_2 into DME over a hybrid CuZnZr/HZSM-5 system in a fixed-bed reactor. On this account, a simulation model was implemented to analyze the inlet and outlet flows related to: (1) a plug-flow DME synthesis reactor (Table 5.1); (2) the liquid/gas separation unit located after the DME reactor (Table 5.2); (3) the distillation column to recover the purified product (Table 5.3).

Being most of the experiments for the direct DME synthesis carried out in the temperature range of 200–300°C and pressure range of 1–5 MPa, for the modeling an operational pressure of 3 MPa at a temperature of 260°C was chosen as a reference condition. The concentration of the outlet gas from DME reactor was calculated starting from a CO_2/H_2 molar ratio of 1:3, by considering a total CO_2 conversion of 30% and a DME selectivity as high as 60%, with a productivity of 1000 kg/year at 0.25 $dm^3/(kg_{cat}{\cdot}s)$ [40]. After reaction, the outlet stream was cooled at 30°C, decompressed at 0.5 MPa and introduced in the separation unit, yielding DME at a purity higher than 99.9% (molar basis) [41].

5.5 OPEN QUESTIONS AND PERSPECTIVES

The feasibility of direct CO_2 utilization on an industrial scale confirms that the widely reported problems associated with slow CO_2 hydrogenation kinetics and catalyst deactivation are surmountable. Key to an effective CCU scenario is the production

TABLE 5.1
Process parameters at the DME reactor (single stage)

Type of process/stream		DME reactor (single stage)	Reaction products from DME reactor	H_2 from electrolyzer	Power input from renewable energy
Start process temperature	°C	260	–	–	–
Start process pressure	MPa	3	–	–	–
Process temperature	°C	260	–	–	–
Process pressure	MPa	3	–	–	–
End process temperature	°C	260	–	–	–
End process pressure	MPa	3	–	–	–
Feed speed	kg/s	2.6×10^{-4}	–	5.5×10^{-5}[a]	
Form of energy supplied	–	electric	–	–	–
Energy requirements	kWh/kg	–	–	–	55[b]
Process kinetics	mol/(kg·s)	0.110	0.095	–	–
		Gas input			
CO_2	mol%	25.00	–	–	–
H_2		75.00	–	–	–
DME		–	–	–	–
MeOH		–	–	–	–
CO		–	–	–	–
H_2O		–	–	–	–
Flow rate	mol/s	0.050	–	0.128	–
Temperature at the inlet unit	°C	200	–	200	–
Pressure at the inlet unit	MPa	3	–	3	–
		Gas output			
CO_2	mol%	–	20.10	–	–
H_2		–	62.65	–	–
DME		–	3.23	–	–
MeOH		–	1.01	–	–
CO		–	1.16	–	–
H_2O		–	11.85	–	–
Flow rate	mol/s	–	0.178	–	–
Temperature at the outlet unit	°C	200	30	–	–
Pressure at the outlet unit	MPa	0.5	0.5	–	–

[a] Mass amount of H_2 produced per unit of time [40].
[b] Electrolytic power consumption as a function of hydrogen flow rate [40].

TABLE 5.2
Process parameters at the liquid/gas separation unit

Type of process		Liquid/gas separation unit	DME/MeOH mix	Gas phase	H_2 separation unit
		Input			
Flow rate	mol/s	0.178	–	–	–
Temperature at the inlet unit	°C	30	–	–	–
Pressure at the inlet unit	MPa	0.5	–	–	–
		Output			
CO_2	s	–	–	23.96	2.00
H_2		–	–	74.65	98.00
DME		–	20.05	–	–
MeOH		–	6.30	–	–
CO		–	–	1.38	–
H_2O		–	73.65		–
Flow rate	mol/s	–	0.030	0.148	0.110[a]
Temperature at the outlet unit	°C	–	30	30	30
Pressure at the outlet unit	MPa	–	0.5	0.5	0.5

[a] H_2 recovered from the gas phase stream.

of renewable hydrogen, which will allow to develop a sustainable, resource-efficient and low-carbon scenario in the next future, through its storage and transportation into liquid organic hydrogen carriers, like chemicals or fuels. In this respect, the handling properties of DME, associated to its ability to be produced from diverse and abundant renewable resources as well as its significant carbon intensity-reducing qualities make it an excellent choice as an alternative clean fuel. The maturity of the available technology for obtaining methanol and the relative simplicity of methanol dehydration are the great attractions of the two-step process to DME. However, the development of competitive processes for its production will play a critical role in helping industry and fleets to meet the new low-carbon fuel/emissions reduction standards around the world. Despite the latest advancements in the formulation of hybrid catalysts to deliver an efficient one-step process of CO_2 hydrogenation to DME, Cu-based systems are expected to remain a reference for CO_2 activation. The main challenges to be dealt with are related to their long-time stability under high-pressure reaction conditions, still requiring novel preparation methodologies (core-shell system, 3D architectures, etc.) to tailor texture, structure, morphology and surface features preventing a fast deactivation. In the same direction, the design of novel reactor configurations for a better heat and water management represents the silver

TABLE 5.3
Process parameters at the distillation column

Type of process		Distillation column	DME	MeOH	Water drain
		Gas input			
Flow rate	mol/s	0.030[a]	–	–	–
Temperature at the inlet unit	°C	260[a]	–	–	–
Pressure at the inlet unit	MPa	0.507[a]	–	–	–
		Gas output			
CO_2	mol%	–	–	–	–
H_2		–	–	–	–
DME		–	98.00	1.00	1.00
MeOH		–	1.00	98.00	1.00
CO		–	–	–	–
H_2O		–	1.00	1.00	98.00
Flow rate	mol/s	–	0.006[a]	0.002[a]	0.022[a]
Temperature at the outlet unit	°C	325[a]	–	–	–
Pressure at the outlet unit	MPa	1.013[a]	–	–	–

[a] From [41].

bullet to establish the technology at a higher readiness level, owing to high CO_2 hydrogenation rates and maximum DME selectivity values.

REFERENCES

[1] Azizi, Z.; Rezaeimanesh, M.; Tohidian, T.; Rahimpour, M.R. Dimethyl ether: A review of technologies and production challenges. *Chemical Engineering and Processing: Process Intensification* **2014**, *82*, 150–172, https://doi.org/10.1016/j.cep.2014.06.007.

[2] Suib, S.L. *New and Future Developments in Catalysis*, Elsevier, Amsterdam, 2013.

[3] Peral, E.; Martín, M. Optimal production of dimethyl ether from switchgrass-based syngas via direct synthesis. *Industrial & Engineering Chemistry Research* **2015**, *54*, 7465–7475, https://doi.org/10.1021/acs.iecr.5b00823.

[4] Steeneveldt, R.; Berger, B.; Torp, T.A. CO_2 Capture and storage: Closing the knowing–doing gap. *Chemical Engineering Research and Design* **2006**, *84*, 739–763, https://doi.org/10.1205/cherd05049.

[5] Fleisch, T.H.; Basu, A.; Sills, R.A. Introduction and advancement of a new clean global fuel: The status of DME developments in China and beyond. *Journal of Natural Gas Science and Engineering* **2012**, *9*, 94–107, https://doi.org/10.1016/j.jngse.2012.05.012.

[6] Bonura, G.; Cannilla, C.; Frusteri, L.; Mezzapica, A.; Frusteri, F. DME production by CO_2 hydrogenation: Key factors affecting the behaviour of CuZnZr/ferrierite catalysts. *Catalysis Today* **2017**, *281*, 337–344, https://doi.org/10.1016/j.cattod.2016.05.057.

[7] Arena, F.; Spadaro, L.; Di Blasi, O.; Bonura, G.; Frusteri, F. Integrated synthesis of dimethylether via CO_2 hydrogenation. *Studies in Surface Science and Catalysis* **2004**, *147*, 385–390, https://doi.org/10.1016/S0167-2991(04)80082-X.

[8] Bonura, G.; Cordaro, M.; Cannilla, C.; Mezzapica, A.; Spadaro, L.; Arena, F.; Frusteri, F. Catalytic behaviour of a bifunctional system for the one step synthesis of DME by CO_2 hydrogenation. *Catalysis Today* **2014**, *228*, 51–57, https://doi.org/10.1016/j.cattod.2013.11.017.

[9] Bonura, G.; Cordaro, M.; Spadaro, L.; Cannilla, C.; Arena, F. Hybrid Cu–ZnO–ZrO_2/H-ZSM5 system for the direct synthesis of DME by CO_2 hydrogenation. *Applied Catalysis B: Environmental* **2013**, *140–141*, 16–24, https://doi.org/10.1016/j.apcatb.2013.03.048.

[10] Frusteri, F.; Cordaro, M.; Cannilla, C.; Bonura, G. Multifunctionality of Cu–ZnO–ZrO_2/H-ZSM5 catalysts for the one-step CO_2-to-DME hydrogenation reaction. *Applied Catalysis B: Environmental* **2015**, *162*, 57–65, https://doi.org/10.1016/j.apcatb.2014.06.035.

[11] Frusteri, F.; Bonura, G.; Cannilla, C.; Drago Ferrante, G.; Aloise, A.; Catizzone, E.; Migliori, M.; Giordano, G. Stepwise tuning of metal-oxide and acid sites of CuZnZr-MFI hybrid catalysts for the direct DME synthesis by CO_2 hydrogenation. *Applied Catalysis B: Environmental* **2015**, *176*, 522–531, https://doi.org/10.1016/j.apcatb.2015.04.032.

[12] Bonura, G.; Cannilla, C.; Frusteri, L.; Frusteri, F. The influence of different promoter oxides on the functionality of hybrid CuZn-ferrierite systems for the production of DME from CO_2-H_2 mixtures. *Applied Catalysis A: General* **2017**, *544*, 21–29, https://-doi.org/10.1016/j.apcata.2017.07.010.

[13] Bonura, G.; Todaro, S.; Frusteri, L.; Majchrzak-Kucęba, I.; Wawrzyńczak, D.; Pászti, Z.; Tálas, E.; Tompos, A.; Ferenc, L.; Solt, H.; Cannilla, C.; Frusteri, F. Inside the reaction mechanism of direct CO_2 conversion to DME over zeolite-based hybrid catalysts. *Applied Catalysis B: Environmental* **2021**, *294*, 120255–120264, https://doi.org/-10.1016/j.apcatb.2021.120255.

[14] Catizzone, E.; Freda, C.; Braccio, G.; Frusteri, F.; Bonura, G. Dimethyl ether as circular hydrogen carrier: Catalytic aspects of hydrogenation/dehydrogenation steps. *Journal of Energy Chemistry* **2021**, *58*, 55–77, https://doi.org/10.1016/j.jechem.2020.09.040.

[15] Catizzone, E.; Bonura, G.; Migliori, M.; Frusteri, F.; Giordano, G. CO_2 recycling to dimethyl ether: State-of-the-art and perspectives. *Molecules* **2018**, *23*, 31–58, https://-doi.org/10.3390/molecules23010031.

[16] Noussan, M.; Raimondi, P.P.; Scita, R.; Hafner, M. The role of green and blue hydrogen in the energy transition—A technological and geopolitical perspective. *Sustainability* **2021**, *13*, 298–323, https://doi.org/10.3390/su13010298.

[17] Hankin, A.; Shah, N. Process exploration and assessment for the production of methanol and dimethyl ether from carbon dioxide and water. *Sustainable Energy & Fuels* **2017**, *1*, 1541–1556, https://doi.org/10.1039/c7se00206h.

[18] Roode-Gutzmer, Q.I.; Kaiser, D.; Bertau, M. Renewable methanol synthesis. *ChemBioEng Reviews* **2019**, *6*, 209–236, https://doi.org/10.1002/cben.201900012.

[19] Dalena, F.; Senatore, A.; Basile, M.; Knani, S.; Basile, A.; Iulianelli, A. Advances in methanol production and utilization, with particular emphasis toward hydrogen generation via membrane reactor technology. *Membranes* **2018**, *8*, 98–125, https://doi.org/10.3390/membranes8040098.

[20] Dang, S.; Qin, B.; Yang, Y.; Wang, H.; Cai, J.; Han, Y.; Li, S.; Gao, P.; Sun, Y. Rationally designed indium oxide catalysts for CO_2 hydrogenation to methanol with high activity and selectivity. *Science Advances* **2020**, *6*, https://doi.org/10.1126/sciadv.aaz2060.

[21] Martin, O.; Martín, A.J.; Mondelli, C.; Mitchell, S.; Segawa, T.F.; Hauert, R.; Drouilly, C.; Curulla-Ferré, D.; Pérez-Ramírez J.; Indium oxide as a superior catalyst for methanol synthesis by CO_2 hydrogenation. *Angewandte Chemie, International Edition* **2016**, *55*, 6261–6265, https://doi.org/10.1002/anie.201600943.

[22] Sun, K.; Fan, Z.; Ye, J.; Yan, J.; Ge, Q.; Li, Y.; He, W.; Yang, W.; Liu, C. Hydrogenation of CO_2 to methanol over In_2O_3 catalyst. *Journal of CO_2 Utilization* **2015**, *12*, 1–6, https://doi.org/10.1016/j.jcou.2015.09.002.

[23] Jiang, X.; Nie, X.; Guo, X.; Song, C.; Chen, J.G. Recent advances in carbon dioxide hydrogenation to methanol via heterogeneous catalysis. *Chemical Reviews* **2020**, *120*, 7984–8034, https://doi.org/10.1021/acs.chemrev.9b00723.

[24] García-Trenco, A.; Valencia, S.; Martínez, A. The impact of zeolite pore structure on the catalytic behavior of CuZnAl/zeolite hybrid catalysts for the direct DME synthesis. *Applied Catalysis A: General* **2013**, *A 468*, 102–111, https://doi.org/10.1016/j.apcata.2013.08.038.

[25] Li, J.L.; Zhang, X.G.; Inui, T. Improvement in the catalyst activity for direct synthesis of dimethyl ether from synthesis gas through enhancing the dispersion of CuO/ZnO/γ-Al_2O_3 in hybrid catalysts. *Applied Catalysis A: General* **1996**, *A 147*, 23–33, https://doi.org/10.1016/S0926-860X(96)00208-6.

[26] Alharbi, W.; Kozhevnikova, E.F.; Kozhevnikov, I.V. Dehydration of methanol to dimethyl ether over heteropoly acid catalysts: The relationship between reaction rate and catalyst acid strength. *ACS Catalysis* **2015**, *5*, 7186–7193, https://doi.org/10.1021/acscatal.5b01911.

[27] Ladera, R.M.; Fierro, J.L.G.; Ojeda, M.; Rojas, S. TiO_2-supported heteropoly acids for low-temperature synthesis of dimethyl ether from methanol. *Journal of Catalysis* **2014**, *312*, 195–203, https://doi.org/10.1016/j.jcat.2014.01.016.

[28] Aloise, A.; Marino, A.; Dalena, F.; Giorgianni, G.; Migliori, M.; Frusteri, L.; Cannilla, C.; Bonura, G.; Frusteri, F.; Giordano, G. Desilicated MFI-type zeolite: Catalytic performances assessment in Methanol to DME dehydration. *Microporous and Mesoporous Material* **2020**, *302*, 110198–110205, https://doi.org/10.1016/j.micromeso.2020.110198.

[29] Migliori, M.; Catizzone, E.; Aloise, A.; Bonura, G.; Gómez-Hortigüela, L.; Frusteri, L.; Cannilla, C.; Frusteri, F.; Giordano, G. New insights about coke deposition in methanol-to-DME reaction over MOR-, MFI- and FER-type zeolites. *Journal of Industrial and Engineering Chemistry* **2018**, *68*, 196–208, https://doi.org/10.1016/j.jiec.2018.07.046.

[30] Bonura, G.; Migliori, M.; Frusteri, L.; Cannilla, C.; Catizzone, E.; Giordano, G.; Frusteri, F. Acidity control of zeolite functionality on activity and stability of hybrid catalysts during DME production via CO_2 hydrogenation. *Journal of CO_2 Utilization* **2018**, *24*, 398–406, https://doi.org/10.1016/j.jcou.2018.01.028.

[31] Liu, R.; Qin, Z.; Ji, H.; Su, T. Synthesis of dimethyl ether from CO_2 and H_2 using a Cu–Fe–Zr/HZSM-5 catalyst system. *Industrial & Engineering Chemistry Research* **2013**, *52*, 16648–16655, https://doi.org/10.1021/ie401763g.

[32] Wang, S.; Mao, D.; Guo, X.; Wu, G.; Lu, G. Dimethyl ether synthesis via CO_2 hydrogenation over CuO–TiO_2–ZrO_2/HZSM-5 bifunctional catalysts. *Catalysis Communication* **2009**, *10*, 1367–1370, https://doi.org/10.1016/j.catcom.2009.02.001.

[33] Zhang, M.-H.; Liu, Z.-M.; Lin, G.-D.; Zhang, H.-B. Pd/CNT-promoted Cusingle bond ZrO_2/HZSM-5 hybrid catalysts for direct synthesis of DME from CO_2/H_2. *Applied Catalysis A: General* **2013**, *A 451*, 28–35, https://doi.org/10.1016/j.apcata.2012.10.038.

[34] Zha, F.; Tian, H.; Yan, J.; Chang, Y. Multi-walled carbon nanotubes as catalyst promoter for dimethyl ether synthesis from CO_2 hydrogenation. *Applied Surface Science* **2013**, *285* (Part B), 945–951, https://doi.org/10.1016/j.apsusc.2013.06.150.

[35] García-Trenco, A.; Martínez, A. A rational strategy for preparing Cu–ZnO/H-ZSM-5 hybrid catalysts with enhanced stability during the one-step conversion of syngas to dimethyl ether (DME). *Applied Catalysis A: General* **2015**, *A 493*, 40–49, https://doi.org/10.1016/j.apcata.2015.01.007.

[36] Lee, S.B.; Cho, W.; Park, D.K.; Yoon, E.S. Simulation of fixed bed reactor for dimethyl ether synthesis. *Korean Journal of Chemical Engineering* **2006**, *23*, 522–530, https://-doi.org/10.1007/bf02706789.

[37] Lewnard, J.J.; Hsiung, T.H.; White, J.F.; Brown, D.M. Single-step synthesis of dimethyl ether in a slurry reactor. *Chemical Engineering Science* **1990**, *45*, 2735–2741, https://doi.org/10.1016/0009-2509(90)80165-B.

[38] Yagi, H.; Ohno, Y.; Inoue, N.; Okuyama, K.; Aoki, S. Slurry phase reactor technology for DME direct synthesis. *International Journal of Chemical Reactor Engineering* **2010**, *8* (1), https://doi.org/10.2202/1542-6580.2267.

[39] Lu, W.-Z.; Teng, L.-H.; Xiao, W.-D. Simulation and experiment study of dimethyl ether synthesis from syngas in a fluidized-bed reactor. *Chemical Engineering Science* **2004**, *59*, 5455–5464, https://doi.org/10.1016/j.ces.2004.07.031.

[40] Ball, M.; Weeda, M. The hydrogen economy – Vision or reality? In *Compendium of Hydrogen Energy*; Woodhead Publishing Series in Energy; Elsevier; **2015**; 237–266.

[41] Bai, Z.; Ma, H.; Zhang, H.; Ying, W.; Fang, D. Process simulation of dimethyl ether synthesis via methanol vapor phase dehydration. *Polish Journal of Chemical Technology* **2013**, *15*, 122–127, https://doi.org/10.2478/pjct-2013-0034.

6 Strategies for carbon capture in concrete production

Rita Nogueira, André Silva, and José Alexandre Bogas

CONTENTS

6.1 INTRODUCTION

Concrete is the most widely used construction material, being considered fundamental for the development of societies around the world. It has the second highest annual volume consumption, only surpassed by water [1, 2]. Despite many advantages (concrete is an inexpensive and versatile building material), its production is associated to a significant environmental impact due to the emission of carbon dioxide (CO_2). In fact, concrete production is responsible for the emission of approximately 8% of the global manmade CO_2 emissions, which is expected to be further aggravated due to the expected 20% increase in cement consumption by 2050 [2, 3].

Ordinary concrete is usually composed of three main components: cement, aggregate and water (aside from chemical and mineral admixtures in minor amount). The mining, processing and transport operations regarding the transformation of raw

DOI: 10.1201/9781003336587-6

material into aggregates consume not negligible amounts of energy, which aggravates the total CO_2 emissions from concrete. However, the major contributing cause for these emissions is the cement manufacturing, accountable for over 80% (by mass) of the total [2, 4, 5].

Focusing on cement production, the main contribution to CO_2 emissions is concentrated in the pyro-processing of raw materials (limestone among others) to obtain clinker, the main binding component in cement. Around 86.5 g of CO_2 is realized per 100 g of clinker, which accounts for 85% of the total [2]. Nearly 60% of this amount is attributed to the thermal decomposition (calcination) of limestone ($CaCO_3$) into calcium oxide (CaO), during cement production [2, 6]. Additionally, the heating in the kiln has to reach 1450°C to promote the formation of cement compounds responsible for their mechanical resistance. This process involves the consumption of a high amount of thermal energy, requiring significant quantities of fuel, in which coal is the dominant source worldwide. Fuel combustion is estimated to account for 35–40% of the CO_2 emissions from cement [4, 7].

Carbon capture, utilization and storage technology (CCUS, previously CCS) is one of the levers to reduce CO_2 emissions in this industry, identified in 2009 by the International Energy Agency (IEA) in Technology Roadmap: Low-Carbon Transition in the Cement Industry. Only recently this strategy was considered to have the potential to be competitive, after external pushes were implemented on governmental level, such as carbon tax or emission trading systems. It is then necessary to explore new solutions where CO_2 can be reused in an efficient way, as will be discussed in this chapter [8, 9].

Cement based-materials show carbonation potential because they contain calcium-bearing compounds, which are carbon reactive. During concrete's service life, these compounds re-absorb atmospheric CO_2 in a process named natural carbonation. This extensively documented process starts at the exposed surface and progressively moves inwards in a slow process, most of the times reported as harmful for reinforced concrete structures due to the depassivation of the reinforcement steel bars. Carbonation is a natural process that, depending on the concrete composition and exposure conditions, is responsible for the absorption of an average estimate of 16% of the concrete-related CO_2 emission during its service life and 1.5% after decommissioning, Figure 6.1. Nevertheless, this only corresponds to around 38% (maximum) of the concrete carbonation potential [4, 7].

Figure 6.2 presents three strategies for carbon capture in concrete production. The first strategy is the accelerated carbonation curing of concrete, whose industrial process is still a long way off, despite some CO_2 capture effectiveness reported. A second strategy consists of incorporating CO_2 during the mixing process, which has been shown to speed up the hardening reactions of cement, although this effect is of short-term and may negatively affect its long-term performance. Finally, the carbonation of cement waste is discussed as a strategy that maximizes the access of CO_2 to the calcium-rich compounds and enables the reuse of concrete as a filler addition, contributing also to the clinker substitution also proposed in the Roadmap. An overview of these three strategies is presented below, discussing the technological features of the carbonated products, the industrial viability of the processes and the net CO_2 balance. The pros and cons will also be addressed, as well as the main findings

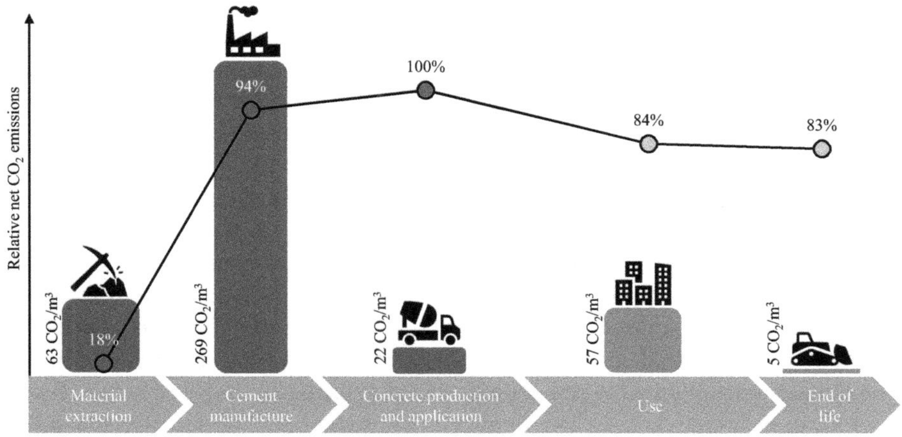

FIGURE 6.1 CO_2 cycle in cement-based materials (case of concrete). Values computed using CO_2 mass per unit volume of concrete.

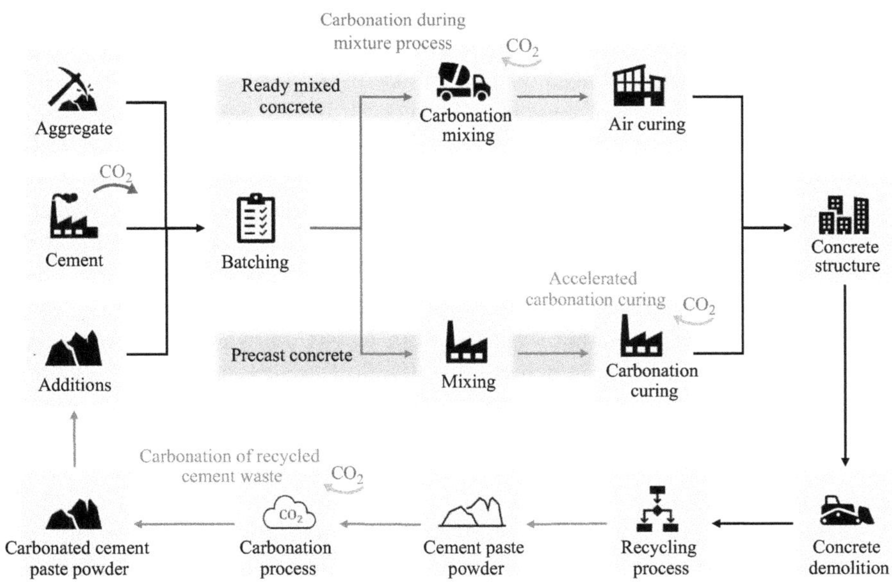

FIGURE 6.2 Strategies for carbon capture in concrete production.

achieved so far, aiming at reducing the carbon footprint of concrete production, the most widely used building material on earth.

6.2 CO_2 CYCLE IN CEMENT-BASED MATERIALS

Although cement corresponds to only 13–18% of ordinary concrete mass, it is a crucial component for the concrete's properties [5]. Cement is a hydraulic binder (ability to set and harden under water) that after reaction with water develops a densely cross-linked matrix of calcium silicate hydrates (C-S-H) and other minor hydration products. Besides concrete, cement can have a diversity of applications in the construction sector, like mortar or grout, comprising the category of cement-based materials.

Dating back to the Industrial Revolution, the manufacturing process of cement has been continuously improved throughout the years, being only standardized in the beginning of the 20th century with the dissemination of Portland cement. Currently, its manufacturing process includes several steps of relative complexity. The cement manufacturing process initiates with the mining of raw materials, generally limestone, clay (or marl) and sand, which are the sources of the calcareous, siliceous and aluminous compounds. These raw materials are grounded, blended in the required proportions and heated to high temperatures (≈1450°C). The clinker thus obtained is grinded up to an average particle size of about 10 μm, in order to become reactive. The final cement product is obtained after the addition of gypsum [10].

Concrete and other cement-based materials are composed of a cement matrix, yielded from the reaction of cement and water, with binding ability to provide cohesion to the loose aggregates (usually sand and gravel) also present. The binding effect is a result from a hydration reaction between cement anhydrous compounds and water. The main products of this reaction are a calcium silicate hydrate gel (C-S-H), with the binding function, and calcium hydroxide (CH) [11].

The hydration kinetics of cement is in general divided into four periods: pre-induction, induction, acceleration and deceleration. The pre-induction period is mainly characterized by the surface dissolution of the anhydrous compounds in reactive sites. This rapid reaction, producing a short-lived hydration with a simultaneous and large heat release, decreases quickly within a few minutes. The beginning of the induction period starts with a drastic slowing down of the rate of hydration, whose reasons are still not well known. After typically 3–4 hours the reactions speed up again, beginning the acceleration period. Both C-S-H and CH grow rapidly, in a process that lasts 4–6 hours in general. Finally, in the deceleration period, there is a reduction of the hydration rate caused by the lack of porous space for new compounds to grow. The hydration process continues for many days, but, for practical reasons, it is usually accessed after 28 days, moment when the hydration degree reaches around 80% [12].

The natural carbonation process starts at the exposed concrete surface and progressively moves inwards in a slow and diffusion-controlled process. The carbonation itself does not cause significant deterioration of concrete (at least for the carbonation degree usually achieved), but has significant effect on reinforced concrete durability by affecting the steel bars embedded inside. The uncarbonated cement paste forms a passive layer of oxide, with high pH (from 12.6 to 13.5), that protects the steel against corrosion, hindering a significant reaction of steel with oxygen and water. Concerning the carbonation reaction, after reacting with other alkaline compounds in pure solution, the CO_2 reacts with the CH responsible for maintaining the high alkalinity, and forms calcium carbonate ($C\bar{C}$), lowering the pH of the liquid phase [8]. Once the pH lowers in the vicinity of the steel bars, the protective oxide film is removed and the rate of corrosion increases significantly [10].

The natural carbonation of concrete is a slow process that is affected by the ability of CO_2 to diffuse through the pore system of the hardened cement paste. If the pores are saturated with water, the diffusion of CO_2 is almost absent since this process is 4 orders of magnitude slower in water than in air. Conversely, if the pores are almost empty of water, the CO_2 remains in the gaseous form unable to react with the hardened cement paste. Thus, the highest rate of carbonation of concrete happens at a relative humidity (RH) of 50–70%. Moreover, the $C\bar{C}$ precipitates from the carbonation reaction occupy a greater volume than CH, thus lowering the porosity of carbonated concrete and hindering the CO_2 diffusion.

Recent studies motivated by the current concerns with the high CO_2 emissions from cement industry, revealed a feasible prospect for CO_2 uptake in concrete, as already overviewed in the introduction. This potential CO_2 uptake relies on the high available reactive calcium content from different compounds existing in cement, during all the different stages of service life, until the demolition of concrete structures and waste treatment (mature cement paste). Industrial processes ought to be developed, considering forced carbonation, also called accelerated carbonation, in contrast with natural carbonation.

6.3 ACCELERATED CARBONATION CURING OF CONCRETE

To be competitive, the precast concrete industry relies on the product turn-over. Hence, the efficiency of the process is focused on the curing stage of concrete and in the implementation of strategies to accelerate the concrete strength development. Commonly, the curing stage is performed using steam (a very energy-intensive process) to obtain an environment of high temperature (up to 70°C) and RH (above 95%). In the 1970s, CO_2 was proposed as an alternative curing source for precast concrete; however, the cost associated to CO_2 production and some reported durability problems hampered the CO_2 utilization on a larger scale. More recently, the interest in accelerated CO_2 curing resurged as a consequence of a combination of factors: cost reduction of CO_2 production; progress in carbon-capture technologies; recent focus of the community on the mitigation of greenhouse gas emissions. In fact, the expectable cost reduction of CO_2 production makes the cost of CO_2 curing similar or even lower than that of the steam curing [13]. Besides the economic issue, the carbon capture potentially provides a positive environmental impact. However, an industrial feasible process has not been established so far and the implementation of this strategy has been addressed essentially by the scientific community [9, 14].

Precast concrete includes dry-mix and wet-mix products. The first category includes most of the small non-structural products, like masonry units, paving stones, cement boards and fibre boards, but also hole core slabs and pre-stressed beams for beam and block floors. The fresh mix is very stiff, with moderate to low cement content and low water-to-cement ratio (w/c). The casting and mechanical compaction by an external force are conducted simultaneously and, in general, no mould is required. This manufacturing process is very cost-efficient but yields products of poor appearance, limiting their application range. The wet-mixes are more fluid in the fresh state, formulated with a higher w/c and/or superplasticizer. They are casted inside a mould, compacted by vibration and demoulded after 24 hours. This process is required for decorative products, like precast concrete walls, and high-strength structural elements (columns, beams, slabs) [15]. The carbonation process, respective impact on concrete performance and net CO_2 balance are covered in the following sections.

6.3.1 Acceleration Carbonation Process

As shown in Figure 6.3, the carbonation curing process is usually composed of three stages. First, in the pre-conditioning stage, the samples are casted and placed at relatively dry curing conditions, generally with a temperature of 20–25°C and a RH of 40–60%. These conditions contrast with those used in traditional curing, where RH is usually higher than 95% to ensure proper cement hydration to accomplish with the performance requirements. In this situation, the concrete pores are saturated and CO_2 gas is unable to penetrate them. Therefore, the pre-conditioning stage allows for a reduction of the pore water content. In fact, this stage is more relevant for wet-mix concrete, considering the higher pore water content. Second, in the carbonation stage, the concrete samples are carbonated in either an enclosed or a flow-through system. These processes are different in terms of energy consumption and carbonation efficiency, but both require environment conditions of 60–70% RH (>90% RH can be also used in special situations) and 20–40°C of temperature for an effective carbonation process. These methods are more efficient the higher and coarser the concrete porosity. Regarding the carbonation atmosphere, pure CO_2 gas is the most commonly used, even though flue gas can be more cost-effective and environmental appealing. The CO_2 diffusion is also influenced by the concentration and partial pressure of CO_2 gas in the environment. Thus, positive atmosphere pressures of 0.1–0.5 MPa and CO_2 concentrations of 10–99.9% have been used [16]. Notice that the CO_2 diffusion is continuously hampered through the carbonation progress, a consequence of the precipitation of carbonates on concrete pores, causing porosity reduction, in a similar mechanism to the natural carbonation process. Finally, in the post-conditioning stage, the concrete samples are cured in moist conditions, usually with a RH higher than 95% or even curing under water immersion. After the water depletion of previous stages, these wet curing conditions enable the hydration of the

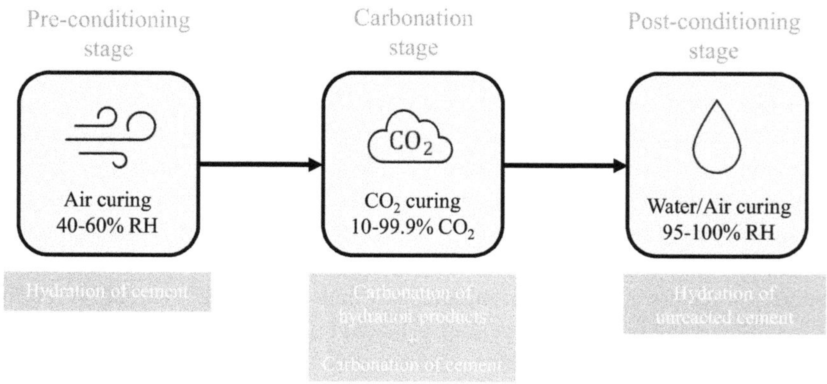

FIGURE 6.3 Carbonation curing stages.

remaining unreacted cement phases, also mitigating the pH reduction of the pore solution [9, 14, 15].

An efficient carbonation curing of concrete is accomplished by ensuring diverse conditions, namely, regarding the microstructure of the cementitious mean. A connected porosity increases the permeability, and consequently the diffusion of CO_2 through the cementitious matrix. Raising the water-to-binder ratio (w/b) of the concrete mixture is one strategy usually used, although limited to w/b = 0.55 [17]. Above this limit the low cement amount in the concrete volume compromises the required carbonation reactions, preventing any further increment of the carbonation degree. The regulation of the pore moisture content during the pre-conditioning stage is also crucial. A maximum CO_2 uptake during the carbonation stage is reported for a water removal during the pre-conditioning stage in the range of 4.5–30% of the initial water content (% of mixing water), depending on the w/b (from 0.34 to 0.5) [14, 18]. To create a sufficient space for CO_2 diffusion, the samples with a higher w/b require less water removal since porous space is higher [14, 15]. Note that a good compromise between hydration and carbonation must be ensured; too harsh drying conditions in the pre-conditioning stage may favour carbonation, but at the expense of hydration, compromising future performance.

Also, the process of forced carbonation has been explored to optimize carbonation efficiency, namely in terms of optimal CO_2 concentration and partial pressure. The carbonation rate exponentially increases for higher CO_2 concentrations, indicating that a higher availability of CO_2 promotes a quicker carbonation reaction and the formation of more reaction products [19]. However, there is a concentration limit for this trend. Experiments showed that the carbonation degree growth is proportionally more significant for a partial pressure of 0.0–0.1 MPa than for 0.1–0.4 MPa, probably because, for this high CO_2 availability, the conditioning factor becomes the concrete permeability to CO_2 (which underwent a stronger reduction due to the higher amount of $C\bar{C}$ precipitates) [20]. A higher duration of the carbonation stage also promotes a higher carbonation degree, with a similar reducing trend over time [13]. Besides these two CO_2-related conditions, the influence of the environment RH and temperature was studied. Studies showed an optimal RH value at around 50% [21]. However, the carbonation degree decrease is relatively low for ranges between RH = 50–65%, reason why the RH at the ambient range is usually applied [16]. Temperature shows a great impact in the carbonation degree. High temperatures (>80°C) accelerate the migration of ions in the liquid phase and promote the CO_2 diffusion; however, it can also overmuch reduce the amount of pore water by evaporation [14]. Regarding the CO_2 binding, the particle size of cement grains is also a crucial factor. The higher specific surface of finer particles promotes a higher carbonation efficiency, given the higher contact area exposed to the carbonated pore water [14, 22]. Table 6.1 presents a summary of results obtained from this carbonation strategy.

6.3.2 Impact on microstructure and performance

The intruding of a carbonation stage in the conventional hydration process may introduce some disturbances with possible impacts in the early-age and latter-age performance and durability of the cement-based materials.

TABLE 6.1
Summary of results from accelerated carbonation curing of cement-based materials

Material	Method	Performance (compressive strength)	CO_2 uptake (wt% of clinker)	Ref.
Wet-mix concrete	PC[a]: 20 h/26°C/50%	+26% at 3 days	14.1%	[23]
(w/b = 0.45)	CS[b]: 6 h/99%/70 kPa after vacuum	+5% at 28 days		
Dry-mix concrete	PC[a]: 0 h	Not evaluated	9.8%	[24]
(w/b = 0.25)	CS[b]: 2 h/99%/507 kPa			
Dry-mix mortar	PC[a]: 0 h	−3.2% NA /5.3% LP	9.8% NA/10.2% LP	[25]
(w/b = 0.15)	PC[a]: 5 h	−3.2% NA/2.1% LP	9% NA/8.5% LP	
No additions (NA)	PC[a]: 11 h	+13.5% NA/19.0% LP	7.9% NA/8% LP	
Addition of LP	PC[a]: 23 h	+12.8% NA/40.2% LP	6.8% NA/6.5% LP	
	PC[a]: 71 h	+37.8% NA/53.4% LP	6.1% NA/5.8% LP	
	CS[b]: 1 h/99%/101 kPa	At 28 days		
Dry-mix paste (w/b = 0.15)	CS[b]: 2 h/1%/101 kPa	+1.2% NA	9.6%	[26]
	CS[b]: 2 h/3%/101 kPa	+3.8% NA	13.2%	
No additions (NA)	CS[b]: 2 h/10%/101 kPa	+3.7% NA	16.6%	
	CS[b]: 2 h/20%/101 kPa	+10.7% NA	19.5%	
		At 3 days		
Wet-mix paste	PC[a]: 10.5 h/25°C/55%	+13.3% NA/12.9% FA	10% NA/15% FA	[27]
(w/b = 0.4)	CS[b]: 2 h/99.8%/507 kPa	+13.3% NA/2.9% FA	17% NA/22% FA	
No additions (NA)	CS[b]: 12 h/99.8%/507 kPa	+22.2% NA/14.3% FA	19% NA/24% FA	
Addition of FA[c]	CS[b]: 24 h/99.8%/507 kPa	At 28 days		
Wet-mix paste/mortar (w/b = 0.4)	PC[a]: 24 h/20°C/60%	+6% NA	9.43% NA	[28]
	CS[b]: 4 h/20%/101 kPa	+17% FA	13.90% FA	
		+39% GGBS	11.60% GGBS	
No additions (NA)		+32% LP	10.40% LP	
Addition of FA[c], GGBS[d], LP[e]		At 3 days		
Wet-mix concrete	PC[a]: 10 h/25°C/55%	+13.4% NA (w/b = 0.3)	11% NA (w/b = 0.3)	[15]
(w/b = 0.3 and w/b = 0.4)	CS[b]: 12 h/99.8%/507 kPa	+10.3% NA (w/b = 0.4)	16.5% NA (w/b = 0.4)	
No additions (NA)		At 28 days		

[a] Pre-conditioning stage: duration/temperature/HR.
[b] Carbonation stage: duration/CO_2 concentration/relative pressure.
[c] Fly ash.
[d] Ground granulated blast-furnace slag.
[e] Lime powder.

Regarding the microstructure, the precipitation of $C\overline{C}$, generally in the form of calcite, promotes a denser microstructure, reducing the total porosity of the cementitious matrix. This impact is function of the pore size: a more effective porosity reduction occurs for the larger pores (>50 nm), with simultaneous increase of nanometre pores (2–7 nm) [29]. However, the subsequent hydration in the post-conditioning stage ensures additional porosity reduction by filling the smaller pores (<50 nm) with hydration products. Thus, the degradation mechanisms supported on transport properties may be hampered, which consequently enhances the material durability. In fact, several studies have reported that carbonation curing of cement-based materials has improved freeze-thaw resistance, chloride penetration resistance and carbonation-induced corrosion resistance [13, 29]. Besides the reduction of CO_2 diffusivity that hinders the risk of reinforcement corrosion, the further hydration in the post-conditioning stage can reset the pH level, which was reduced during the carbonation stage, to a value similar to the conventional concrete curing [14, 15].

The carbonation stage can also partially offset the deceleration of reactions consequence of the water depletion, by two main mechanisms. First, the presence of CO_2 in a freshly added water mixture promotes a faster dissolution of the anhydrous compounds, thus accelerating the hydration rate [25]. The second mechanism regards the nucleation effect of the precipitated $C\overline{C}$, which acts as a site for precipitation and accelerating growth of the reaction products [30]. Hence, compared with conventional hydrated cement, the carbonation curing potentially generates more reaction products, with a reduced porosity. Moreover, the intermix between C-S-H, from the hydration reaction, and $C\overline{C}$, from the carbonation reaction, promotes a higher toughness and hardness, given the strong bonding between these two compounds [30]. Therefore, after carbonation curing the cement-based material shows a widely demonstrated accelerated strength development and microstructure densification. The early age compressive strength increase of the carbonation curing samples is also found at later ages, but it is usually negligible [14, 16].

6.3.3 Influence of the process on the net CO_2 balance

The CO_2 uptake of the carbonation curing process depends heavily on both the conditions of the carbonation stage and the chemical and microstructural characteristics of the cement-based material under curing. Thus, a range between about 20 and 50 kg of CO_2/m^3 of concrete can be considered consensual, taking into account the experiments carried out from diverse authors [23–26].

Despite this technology being limited to the precast concrete industry, it accounts for 20–30% of the total concrete industry, including dry and wet-mix products, thus unravelling a feasible prospect for the CO_2 uptake method that hampers the CO_2 emissions associated with the cement industry. In fact, the energy consumption of carbonation curing is about 20% of that of steam curing [13]. Moreover, if flue gas is considered as the CO_2 gas source along with a natural pre-conditioning stage, the energetic cost can be reduced. This also reduces the CO_2 emissions throughout the curing process, which improves the net CO_2 besides that obtained from the CO_2 uptake. Even though flue gas has a lower CO_2 concentration (15–20%), which decreases the carbonation efficiency, the energy-cost of CO_2 capture, storage and transport would be reduced [9, 16].

6.4 CARBONATION OF CONCRETE DURING MIXING

Despite the promising theoretical considerations in favour of the previous strategy, its application as an efficient sink for carbon lacks demonstration and the main reason argued for is the auto-blocking effect in CO_2 diffusion yielded from carbonation of external layers of the material. In this sense, the carbonation during the mixing process appears to overcome this issue, since cement particles dispersed in water ensure full access of CO_2 to the mean. There are also more practical reasons that favour this strategy, such as the extension of its application to in situ concrete structures instead of only precast concrete products. Next sections discuss this carbonation process, impact on concrete performance and CO_2 uptake.

6.4.1 ACCELERATION CARBONATION PROCESS

The process comprises the addition of CO_2 as a mixture ingredient. Hence, all the reactive compounds: cement, water and CO_2 are put in contact at the same time. Diverse methods have been tested to materialize this strategy. The utilization of previously prepared carbonated water was developed by Kwasny et al. [31] and Silva et al. [32]; then, the solid materials were introduced and mixture proceeded as normal. The expected carbon uptake shall be small, due to the low solubility of CO_2 in water under atmospheric pressure (0.0015 gCO_2/gH_2O [33]). The production of the mixture inside a carbonation chamber with a higher CO_2 concentration was tested by Nogueira et al. [34] and Kwasny et al. [31], the later also adopting carbonated water at the same time. Atmospheric pressure and concentrations of CO_2 around 80–90% v/v, close to the maximum possible, are usually used. But probably the most known process corresponds to the injection of CO_2 into the cement mixture, during mixing. This technology has been used for decades by the industry of wood-fibre boars, to increment the product turn-over [35]. More recently CarbonCure Technology Inc., which markets ready-mixed concrete, implemented a process where CO_2 (mix of gas and snow) is injected into the truck during concrete mixing. This injection lasts a few minutes (1–2 min) and the amount of introduced CO_2 is reported to be lower than 1.0% by weight of cement [36].

In contrast to carbonation curing discussed in Section 6.3, the different processes for carbonation during mixing have not yet been significantly explored, since research on the subject is recent and some setbacks related to the performance reduction have arisen. Research has shown that keeping the CO_2 amount below a given value can avoid this performance reduction. Literature shows that the best way to control this amount is to introduce liquid CO_2 inside the cement mixture for a short period of time [37]. CarbonCure Technology Inc. injects a precise dosage of CO_2 into the concrete during mixing. Concrete presented an increase in compressive strength of 10% at 28 days. This strength increase can be used to reduce the clinker amount in 7%, reducing even more the CO_2 footprint. The concrete did not show flash setting nor workability problems. Other properties such as reduction in pH or density were

TABLE 6.2
Summary of results from carbonation of cement-based materials during mixing

Material	Carbonation process	Performance	Footprint benefit	Ref.
Ready-mix concrete (308 kg of cement and 77 kg of slag)	0.05 wt% 0.15 wt% 0.30 wt% (CO_2/cement) injected during mixing	3% increment 4% reduction 6% reduction In *fc*[a] at 28 days No impact on durability	Not mentioned	[37]
Ready-mix concrete (147.7 kg of cement and 73.9 kg of slag and of fly ash)	0.11 wt% (CO_2/cement) injected during mixing	No reduction in *fc*[a] in relation to a reference concrete with 4.3% more binder	Net reduction in CO_2 of 10.0 kg/m^3 of concrete	[38]
Cement paste (w/c = 0.4)	Carbonated water with 4.2 pH (CO_2 introduced at a pressure of 0.8 MPa)	20% reduction In *fc*[a] at 28 days	Not mentioned	[39]
Cement mortar (w/c = 0.4)	Carbonated water with 4.2 pH (CO_2 introduced at a pressure of 0.8 MPa)	Reduction in *fc* Porosity doubled and pores around 80 nm increased	Not mentioned	[39]
Cement paste (w/c = 0.5)	Carbonated water with 4.2 pH (CO_2 introduced at a pressure of 0.1 MPa)	6% reduction in *fc*[a] at 28 days	Not mentioned	[32]
Cement paste (w/c = 0.44)	45 min 90 min of mixing time in a carbonation chamber with 85 ± 5 vol% of CO_2 concentration	12% reduction 13% reduction In *fc*[a] at 28 days	0.93 wt% 1.12 wt% CO_2 uptake (CO_2/clinker)	[34]

[a] Compressive strength.

also not affected [36]. Table 6.2 presents a summary of results obtained from this carbonation strategy.

6.4.2 Impact on Microstructure and Performance

For CO_2 content higher than a certain amount, the material's performance is worst, either in fresh or in hardened state. Flash setting impairs workability during application. Additionally, a reduction in the mechanical strength was reported [32, 34, 37]. Cement pastes produced in the presence of CO_2 show a looser and more disconnected microstructure in relation to the reference. SEM images of 3 day-pastes also revealed larger zones of well-developed phases of portlandite (CH). Considering that in ordinary pastes C-S-H compounds occur intermixed with CH, it was theorized by Nogueira et al. [34] that CO_2 produces a rather heterogeneous paste where larger portlandite crystals arise intercalated with C-S-H zones, which could explain the worst mechanical strength [40]. Porosimetry results also support an alteration in pore size distribution of the cement paste, coarsening the pore sizes from a unimodal distribution around 0.1 μm, to a widened distribution from 0.1 μm to 1 μm, due to the introduction of CO_2, keeping the total porosity similar [34].

6.4.3 Influence of the Process on the Net CO_2 Balance

Given the negative impact of CO_2 during the mixing process, this strategy cannot take the best advantage of carbon-reactive potential of cement-based materials. Their performance is very much based on the hydration reactions that occur from the first minutes after water addition, and CO_2 interferes negatively with these reactions if added in dosages as low as 1% of the cement weight. It is possible to avoid the described effect if CO_2 is added in a very low amount (below 1%). In this case, increase in mechanical strength can be obtained, as a consequence of the very controlled set-accelerator effect of CO_2. This mechanical strength increase can be offset by cement reduction, which eventually leads to a higher CO_2 reduction in terms of unit volume of concrete. A 14.8 kg of CO_2 saved per cubic meter of concrete is reported in the study from Monkman et al. [36, 37].

6.5 VALORIZATION OF RECYCLED CEMENT WASTE THROUGH CARBONATION

The construction industry is the largest producer of waste in the European Union, accounting for about 35% [41] up to 46% [42] of the total waste stream generation. It is reported that this sector is responsible for about 450 [43] up to 850 [41] million tonnes of construction and demolition waste per year (CDW). CDW comprehends the waste generated by all the activities, involving construction, maintenance, demolition and deconstruction of buildings and civil works. Thus, the proportion of concrete varies significantly, being estimated at values between 12 and 40% [42].

The most common outlet for CDW is dumping and landfilling, which produces environmental problems such as soil and water pollution, given the hazardous substances in presence. Thus, recycling technologies for CDW have been increasingly studied and developed. Currently, the most feasible method for CDW disposal is its conversion into recycled concrete aggregate (RCA) [44]. During the production of RCA, waste concrete is crushed, yielding 10–20% of a waste powder by-product. Although this waste powder is not entirely composed by a hydrated cement paste, having aggregate impurities in its constitution, this powder material is commonly referred as cement paste powder (CPP) [45].

Recent studies have explored the utilization of CPP as a cement substitute, in a role similar to the limestone filer addition, and a positive performance was observed in a cement mortar for a maximum of 5% replacement. With the same purpose, other papers addressed the carbonation potential of CPP prior to its application as an addition. Lu et al. [45] investigated the impact of adopting fully carbonated CPP (CCPP) and uncarbonated CPP on the compressive strength (*fc*) of cement pastes. In this study, the application of CCPP produced better results than those obtained for CPP. *fc* was 6% and 32% higher for an incorporation of 10% and 20%, respectively. Also, Mehdizadeh et al. [46] reported that the incorporation of 5–20% of CCPP returned a positive increase in *fc* when compared with a reference paste, composed of 100% cement, thus uncovering a feasible prospect for CO_2 capture technology [47].

6.5.1 Acceleration carbonation process

The CPP obtained from the recycling facilities is a rather complex and variable material. Hence, academic works usually resort to synthetic laboratory-made cement pastes, with a controlled composition and hydration degree. Generally, after a curing period of about 28–90 days, the hardened pastes are crushed and sieved prior to the carbonation process, into a particle size usually below 150 μm [45, 46, 48]. The finer the particles, the higher the carbonation efficiency of the products. CPP is then placed in a carbonation chamber at RH of 60–70%, a temperature of about 20°C and a CO_2 concentration from 20 to 100%. Wu et al. [48] reported that CPP rapidly reacts with the CO_2; however, after this initial boost, the carbonation rate slows down significantly, over-prolonging the process duration. Thus, the authors settle for a carbonation duration of 3 days. Mehdizadeh et al. [46] extended even more the carbonation period for 28 days. Table 6.3 presents a summary of results obtained from works on this carbonation strategy.

6.5.2 Impact on microstructure and performance

The few works on the application of CCPP indicate a positive contribution of this material as a binder addition in cement-based pastes, when compared with both the 100%-cement paste and the paste with the same CPP replacement. However, in the first case (100%-cement) this benefit was limited to a maximum replacement of 20%. In comparison to CPP, CCPP still showed a positive contribution even for higher replacement percentages CPP [45, 46]. In fact, Wu et al. [48] reported that a 30% addition of CCPP in a cement mortar increased the compressive strength by 12.6% comparing with a mortar with the same addition of CPP. Moreover, the water transport properties were similar to those of a 100%-cement mortar. The positive impact was associated with the presence of $C\bar{C}$ in CCPP, which provides the effects assigned to limestone additions used in blended cements, namely the heterogeneous nucleation effect, responsible for both an increase in nucleation and growth sites for hydration products of cement and a spacing of anhydrous grains, avoiding growth blocking of the new compounds [45, 46, 48].

6.5.3 Influence of the process on the net CO_2 balance

Studies on the environmental analysis of CCPP application at a larger scale production are unknown, since the investigation of this strategy is very recent and limited

TABLE 6.3
Summary of results from accelerated carbonation of recycled cement waste

Cement paste	CPP	Carbonation process[a]	Performance (compressive strength)	CO_2 uptake (wt% of CPP)	Ref.
w/b = 0.4	75 µm/10% 75 µm/20% 75 µm/30%	– 99% 101.3 kPa 20°C 60%	+16%/+4% +24%/+12% −4%/−5% At 24 h/28 days	24.3	[45]
w/b = 0.3	75 µm/150 µm/10% 75 µm/150 µm/15% 75 µm/150 µm/20%	28 days 20% 101.3 kPa 20°C 65%	+5%/0% +8%/0% 0%/−3% At 28 days	21.2	[46]
w/b = 0.3	30%	12 days 20% 101.3 kPa 20°C 70%	–	19.4	[49]

[a] Duration/CO_2 concentration/CO_2 pressure/temperature/HR.

to the scientific community. Van der Zee et al. [50] estimated that 0.056 t of CO_2 can be sequestered by waste concrete, considering that cement typically accounts for 13–18 wt% of concrete. However, recent studies have reported a CO_2 uptake for completely carbonated CCPP of about 0.24–0.28 g of CO_2/g of CCPP, which, if 15% of cement per total waste concrete is considered, corresponds to only 0.036–0.042 g of CO_2/g of waste concrete [45, 49, 51]. Nonetheless, besides the CO_2 uptake by the CCPP addition, also the clinker replacement can give a contribution to reduce the CO_2 emissions [52]. Thus, considering the averaging CO_2 emissions of 0.865 g of CO_2/g of clinker [51], and a 20% replacement of clinker with CCPP, this methodology is able to reduce the clinker-related CO_2 emissions to 0.69 g of CO_2/g of clinker, a reduction of 20.2% [51].

6.6 CARBON CAPTURE STRATEGY FOR CASE STUDY

The three strategies here addressed for obtaining CCUS concrete products (and other cement-based materials) were selected because of the attention they have been receiving by academy and industry for the last decade (carbonation curing) or the high volume of the market in which they have the potential to be applied (carbonation mixing and valorization of recycled cement waste). The pros and cons of the three strategies are summarized in Figure 6.4.

Carbon curing of concrete was the carbonation strategy selected to apply in the case study of the carbon chain in carbon dioxide industrial utilization technologies,

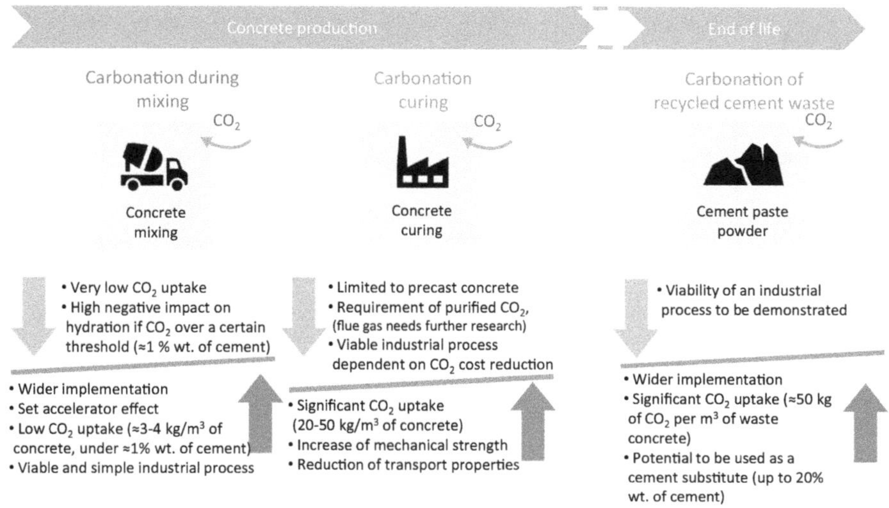

FIGURE 6.4 Comparison of accelerated carbonation strategies.

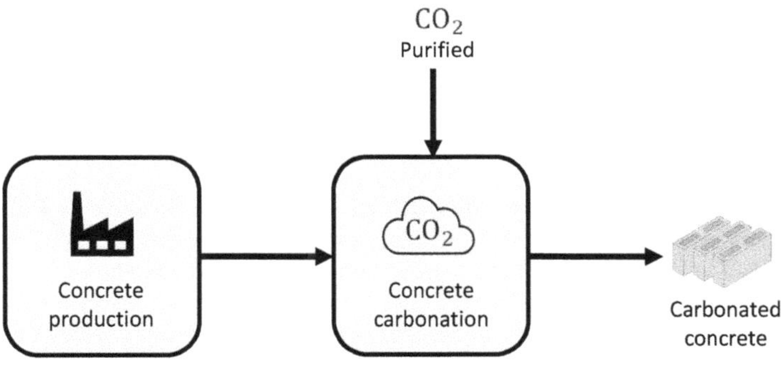

FIGURE 6.5 Scheme of the concrete carbonation process.

Figure 6.5. The main pros that highlight this strategy among the three was the highest CO_2 uptake (values ranging on average from 20 to 50 kg/m^3 of concrete), well supported in a significant amount of literature results. Moreover, the lack of industrial viability reported can be minimized, considering the increment in concrete performance and, especially, the synergy that arises when placing diverse carbon industrial utilization technologies as happens in this case study. Table 6.4 summarizes the main parameters used in the case study.

Parameters presented in Table 6.4 are based on the process implemented by He et al. [24]. Three dry-mix concrete samples with a volume of 0.24 dm^3 each and a density of 2242 kg/m^3 were placed inside a carbonation chamber with a volume of 5.5 dm^3 during 2 hours, right after demoulding. The chamber was able to keep pressure at 0.5 MPa, reached taking advantage of the pressure of the supplied gas (8.3 MPa). The input of the diverse concrete components (cement, aggregates and water)

TABLE 6.4
Concrete carbonation process – parameters used in the case study

Parameter	Unit	Value
Cement input	kg/kg concrete	0.20
	kg/h	0.15
Aggregates and water input	kg/kg concrete	0.80
	kg/h	0.66
CO_2 absorption coefficient	kg/kg concrete	0.02
	kg/h	~0.02
Concrete input	kg/h	0.81
Carbonated concrete output	kg/h	0.83
Process temperature	°C	25
Process pressure	MPa	0.5
Process duration	H	2
Gas composition	–	100% CO_2
Energy for cement production [53]	kWh/tonne cement	117
CO_2 coefficient emission [51]	kg CO_2/tonne cement	865

is computed from the cement amount in the mixture and the CO_2 absorption coefficient was based on the values measured by the authors after the carbonation process.

REFERENCES

[1] International Energy Agency; World Business Council for Sustainable Development. *Cement Technology Roadmap 2009.* International Energy Agency, 2009. ISBN: 978-3-940388-47-6, https://www.wbcsd.org/contentwbc/download/4586/61682/1

[2] Filippo, J.; Karpman, J.; DeShazo, J. The impacts of policies to reduce CO_2 emission within the concrete supply chain. *Cement and Concrete Composites* **2019**, *101*, 67–82, https://doi.org/10.1016/j.cemconcomp.2018.08.003.

[3] Environment, U.; Scrivener, K.; John, V.; Gartner E. Eco-efficient cements: Potential economically viable solutions for a low CO_2 cement-based materials industry. *Cement and Concrete Research* **2018**, *114*, 2–26, https://doi.org/10.1016/j.cemconres.2018.03.015.

[4] Lippiatt, N.; Ling, T.; Pan, S. Towards carbon-neutral construction materials: Carbonation of cement-based materials and the future perspective. *Journal of Building Engineering* **2020**, *28*, 101062, https://doi.org/10.1016/j.jobe.2019.101062.

[5] Metha, P. Reducing the environmental impact of concrete: concrete can be durable and environmentally friendly. *Concrete International* **2001**, *10*, 61–66, www.maquinamole.net/EcoSmartConcrete.com/docs/trmehta01.pdf.

[6] Carriço, A.; Bogas, A.; Guedes, M. Thermoactivated cementitious materials – A review. *Construction and Building Materials* **2020**, *250*, 118873, https://doi.org/10.1016/j.conbuildmat.2020.118873.

[7] Pade, C.; Guimaraes, M. The CO_2 uptake of concrete in a 100 year perspective. *Cement and Concrete Research* **2007**, *37*, 1348–1356, https://doi.org/10.1016/j.cemconres.2007.06.009.

[8] Jang, J.; Kim, G.; Kim, H.; Lee, H. Review on recent advances in CO_2 utilization and sequestration technologies in cement-based materials. *Construction and Building Materials* **2016**, *127*, 762–773, https://doi.org/10.1016/j.conbuildmat.2016.10.017.

[9] Zhang, D.; Ghouleh, Z.; Shao, Y. Review on carbonation curing of cement-based materials. *Journal of CO_2 Utilization* **2017**, *21*, 119–131, https://doi.org/10.1016/j.jcou.2017.07.003.
[10] Neville A.; *Properties of Concrete*, 5th ed., Prentice Hall, 2012, p. 872.
[11] Beaudoin, J.; Odler, I. 5-Hydration, setting and hardening of Portland cement. In *Lea's Chemistry of Cement and Concrete*, 5th ed., Elsevier Ltd, 2019, 157–242, https://doi.org/10.1016/B978-0-08-100773-0.00005-8.
[12] Scrivener, K.; Ouzia, A.; Juilland, P.; Mohamed, A. Advances in understanding cement hydration mechanisms. *Cement and Concrete Research* **2019**, *124*, 105823, https://doi.org/10.1016/j.cemconres.2019.105823.
[13] Rostami, V.; Shao, Y.; Boyd, A. Carbonation curing versus steam curing for precast concrete production. *Journal of Materials in Civil Engineering* **2012**, *24*, 1221–1229, doi/full/10.1061/(ASCE)MT.1943-5533.0000462.
[14] Liu, Z.; Meng, W. Fundamental understanding of carbonation curing and durability of carbonation-cured cement-based composites: A review. *Journal of CO_2 Utilization* **2021**, *44*, 101428, https://doi.org/10.1016/j.jcou.2020.101428.
[15] Zhang, D.; Shao, Y. Early age carbonation curing for precast reinforced concretes. *Construction and Building Materials* **2016**, *113*, 134–143, https://doi.org/10.1016/j.conbuildmat.2016.03.048.
[16] Liu, B.; Qin, J.; Shi, J.; Jiang, J.; Wu, X.; He, Z. New perspectives on utilization of CO_2 sequestration technologies in cement-based materials. *Construction and Building Materials* **2021**, *272*, 121660, https://doi.org/10.1016/j.conbuildmat.2020.121660.
[17] Wang, J.; Xu, H.; Xu, D.; Du, P.; Zonghui, Z.; Yuan, L.; Cheng, X. Accelerated carbonation of hardened cement pastes: Influence of porosity. *Construction and Building Materials* **2019**, *225*, 159–169, https://doi.org/10.1016/j.conbuildmat.2019.07.088.
[18] Shi, C.; He, F.; Wu, Y. Effect of pre-conditioning on CO_2 curing of lightweight concrete blocks mixtures. *Construction and Building Materials* **2012**, *26*, 257–267, https://doi.org/10.1016/j.conbuildmat.2011.06.020.
[19] Shi, C.; Liu, M.; He, P.; Ou, Z. Factors affecting kinetics of CO_2 curing of concrete. *Journal of Sustainable Cement-Based Materials* **2012**, *1*, 24–33, https://doi.org/10.1080/21650373.2012.727321.
[20] Lagerblad, B. Carbon dioxide uptake during concrete life cycle - state of the art. *Cement och Betong Institutet* **2007**.
[21] Xuan, D.; Zhan, B.; Poon, C. A maturity approach to estimate compressive strength development of CO_2-cured concrete blocks. *Cement and Concrete Composites* **2018**, *85*, 153–160, https://doi.org/10.1016/j.cemconcomp.2017.10.005.
[22] Polettini, A.; Pomi, R.; Stramazzo, A. Carbon sequestration through accelerated carbonation of BOF slag: Influence of particle size characteristics. *Chemical Engineering Journal* **2016**, *298*, 26–35, https://doi.org/10.1016/j.cej.2016.04.015.
[23] Sharma, D.; Goyal, S. Effect of accelerated carbonation curing on near surface properties of concrete. *European Journal of Environmental and Civil Engineering* **2020**, 1964–8189 (Print), 2116–7214 (Online), https://doi.org/10.1080/19648189.2019.1707714.
[24] He, Z.; Wang, S.; Mahoutian, M.; Shao, Y. Flue gas carbonation of cement-based building products. *Journal CO_2 of Utilization* **2020**, *37*, 309–319, https://doi.org/10.17632/9nvzwmdz8g.1.
[25] Zhang, D.; Cai, X.; Jaworska, B. Effect of pre-carbonation hydration on long-term hydration of carbonation-cured cement-based materials. *Construction and Building Materials* **2020**, *231*, 117122, 10.1016/j.conbuildmat.2019.117122.
[26] Li, X.; Ling, T. Instant CO_2 curing for dry-mix pressed cement pastes: Consideration of CO_2 concentrations coupled with further water curing. *Journal of CO_2 Utilization* **2020**, *37*, 348–354, https://doi.org/10.1016/j.jcou.2020.01.001.
[27] Zhang, D.; Cai, X.; Shao, Y. Carbonation curing of precast fly ash concrete. *Journal of Materials in Civil Engineering* **2016**, *28*, https://doi.org/10.1061/(ASCE)MT.1943-5533.0001649.

[28] Qin, L.; Gao, X.; Chen, T. Influence of mineral admixtures on carbonation curing of cement paste. *Construction and Building Materials* **2019**, *212*, 653–662, https://doi.org/10.1016/j.conbuildmat.2019.04.033.

[29] Zhang, D.; Shao, Y. Surface scaling of CO_2-cured concrete exposed to freeze-thaw cycles. *Journal of CO_2 Utilization* **2018**, *27*, 137–144, https://doi.org/10.1016/j.jcou.2018.07.012.

[30] Ouyang, X.; Koleva, D.; Ye, G.; Breugel, K. Understanding the adhesion mechanisms between C-S-H and fillers. *Cement and Concrete Research* **2017**, *100*, 275–283, https://doi.org/10.1016/j.cemconres.2017.07.006.

[31] Kwasny, J.; Muhammed Basheer, P.; Russell, M. *CO_2 Sequestration in Cement-Based Materials during Mixing Process Using Carbonated Water and Gaseous CO_2*. 4th International Conference on the Durability of Concrete Structures, 2014.

[32] Silva, A.; Nogueira, R.; Bogas, A.; Rodrigues, M. *Influence of Carbon Dioxide as a Mixture Component on the Cement Hydration*. The 4th International RILEM Conference, Delft, The Netherlands, 2021.

[33] Dodds, W.; Stutzman, L.; Sollami, B. Carbon dioxide solubility in water. *Industrial and Engineering Chemistry* **1956**, *1*, 92–95, https://doi.org/10.1021/i460001a018.

[34] Nogueira, R.; Silva, A.; Bogas, G.; Gomes, J. Early-age hydration of Portland cement under forced carbonation. To be submitted in October 2022.

[35] Jorge, F.; Pereira, C.; Ferreira, J. M. Wood-cement composites: A review. *Holz Roh Werkst* **2004**, *62*, 370–377, https://doi.org/10.1007/s00107-004-0501-2.

[36] CarbonCure Home Page, https://www.carboncure.com/technology/ (Accessed June 07, 2021), CarbonCure Technologies, 2021.

[37] Monkman, S.; MacDonald, M.; Doug Hooton, R.; Sandberg, P. Properties and durability of concrete produced using CO_2 as an accelerating admixture. *Cement and Concrete Composites* **2016**, *74*, 218–224, https://doi.org/10.1016/j.cemconcomp.2016.10.007.

[38] Monkman, S. *Sustainable Ready Mixed Concrete Production Using Waste CO_2: A Case Study*. Recent Advances in Concrete Technology and Sustainability Issues: Fourteenth International Conference, Beijing, China, 2018.

[39] Lippiatt, N.; Ling, T. Rapid hydration mechanism of carbonic acid and cement. *Journal of Building Engineering* **2020**, *31*, 101357, https://doi.org/10.1016/j.jobe.2020.101357.

[40] Cuesta, A.; Zea-Garcia, J.; Londono-Zuluaga, D.; Torre, A.; Santacruz, I.; Vallcorba, O.; Dapiaggi, M.; Sanfélix, S.; Aranda, M. Multiscale understanding of tricalcium silicate hydration reactions. *Scientific Reports* **2018**, *8*, 8544, https://doi.org/10.1038/s41598-018-26943-y.

[41] Sáez, P.; Osmani, M. A diagnosis of construction and demolition waste generation and recovery practice in the European Union. *Journal of Cleaner Production* **2019**, *241*, 188400, https://doi.org/10.1016/j.jclepro.2019.118400.

[42] Gálvez-Martos, J.; Styles, D.; Schoenberger, H.; Zeschmar-Lahl, B. Construction and demolition waste best management practice in Europe. *Resources, Conservation and Recycling* **2018**, *136*, 166–178, https://doi.org/10.1016/j.resconrec.2018.04.016.

[43] Cembureau, *Cement, Concrete & The Circular Economy*. Cembureau, European Cement Association, 2016.

[44] Tang, Q.; Ma, Z.; Wu, H.; Wang, W. The utilization of eco-friendly recycled powder from concrete and brick waste in new concrete: A critical review. *Cement and Concrete Composites* **2020**, *114*, 103807, https://doi.org/10.1016/j.cemconcomp.2020.103807.

[45] Lu, B.; Shi, C.; Zhang, J.; Wang, J. Effects of carbonated hardened cement paste powder on hydration and microstructure of Portland cement. *Construction and Building Materials* **2018**, *186*, 699–708, https://doi.org/10.1016/j.conbuildmat.2018.07.159.

[46] Mehdizadeh, H.; Ling, T.; Cheng, X.; Mo, K. Effect of particle size and CO_2 treatment of waste cement powder on properties of cement paste. *Canadian Journal of Civil Engineering* **2020**, *48*, 522–531, https://doi.org/10.1139/cjce-2019-0574.

[47] Katsuyama, Y.; Yamasaki, A.; Iizuka, A.; Fujii, M.; Kumagai, K.; Yanagisawa, Y. Development of a process for producing high-purity calcium carbonate ($CaCO_3$) from waste cement using pressurized CO_2. *Environmental Progress* **2005**, *24*, 162–170, https://doi.org/10.1002/ep.10080.

[48] Wu, H.; Liang, C.; Xiao, J.; Ma, Z. Properties and CO_2-curing enhancement of cement-based materials containing various sources of waste hardened cement paste powder. *Journal of Building Engineering* **2021**, *44*, 102677, https://doi.org/10.1016/j.jobe.2021.102677.

[49] Ouyang, X.; Wang, L.; Xu, S.; Ma, Y.; Ye, G. Surface characterization of carbonated recycled concrete fines and its effect on the rheology, hydration and strength development of cement paste. *Cement and Concrete Composites* **2020**, *114*, 103809, https://doi.org/10.1016/j.cemconcomp.2020.103809.

[50] Van der Zee, S.; Zeman, F. Production of carbon negative precipitated calcium carbonate from waste concrete. *Canadian Journal of Chemical Engineering* **2016**, *94*, 2153–2159, https://doi.org/10.1002/cjce.22619.

[51] Skocek, J.; Zajac, M.; Haha, M. Carbon Capture and Utilization by mineralization of cement pastes derived from recycled concrete. *Nature Research* **2020**, *10*, 5614, https://doi.org/10.1038/s41598-020-62503-z.

[52] Bouarroudj, M.; Rémond, S.; Bulteel, D.; Potier, G.; Michel, F.; Zhao, Z.; Courard, L. Use of grinded hardened cement pastes as mineral addition for mortars. *Journal of Building Engineering* **2021**, *34*, 101863, https://doi.org/10.1016/j.jobe.2020.101863.

[53] Global Cement and Concrete Association Home Page, https://gccassociation.org/ (Accessed September 13, 2021), GNR Project Reporting CO_2.

7 Methodologies for CCU technologies environmental analysis

Marcello De Falco, Gianluca Natrella, Paulina Popielak, and Mauro Capocelli

CONTENTS

7.1 INTRODUCTION

Carbon capture and utilization (CCU) represents a smart solution to re-use CO_2 as a reagent for the synthesis of high-end products instead of storing it underground, implementing a "Circular Economy" concept [1]. Many technologies in this field

DOI: 10.1201/9781003336587-7

are recently attracting attention [2–6], but CO_2 is a very stable compound and its utilization as a reagent is energetic intensive. As a consequence, it is unclear whether CCU processes allow for a net reduction of environmental impacts from a life cycle perspective and whether these solutions are sustainable.

In this chapter, we analyze a complete process architecture able to convert CO_2 in dimethyl ether (DME, IUPAC name methoxymethane), an organic compound mainly used as a reagent for the synthesis of widely applied products such as dimethyl sulfate (a methylating agent), methyl acetate and light olefins. In addition, a growing interest on DME synthesis is due to its characteristics similar to those of liquefied petroleum gas (LPG) and to the excellent combustion property that make the DME a potential candidate as a fuel in compression-ignition engines [7]. Adding DME to the diesel fuel allows an enhancement of the combustion properties, mainly in terms of reduction of smoke, NO_x and CO_2 emissions [8].

However, even if CCU is the best solution to treat CO_2 theoretically, in practical terms the need to build a chemical processing plant involves significant energy consumption and additional carbon dioxide emissions (beyond a significant Capital Expenditure). For this reason, there is the need to analyze in detail the pros and cons of these innovative configurations in terms of energy expenditure and environmental impacts.

Some strategies for process analysis, focused on the quantification of the performance, energy and exergy efficiency, carbon footprints, are briefly reported and described. Then, a process scheme for the sustainable production of DME from pure CO_2 and green hydrogen (i.e. hydrogen produced through electrolysis using 100% renewable electricity) is described, modeled and simulated.

7.2 PROCESS SIMULATION AND ENVIRONMENTAL ANALYSES

There are many tools to evaluate the performance of a process configuration. In this paragraph, we focus the attention on three of them:

- Process simulation by means of process analysis software.
- Exergy balance and efficiency.
- Life Cycle Assessment (LCA).

Each of this allows to evaluate the behavior of a plant from a different point of view:

- The process simulation quantifies the performance of the plant and of each unit composing it, the input/output streams and their physical characteristics (pressure, temperature, flowrate, composition). These tools allow to optimize the performance of a process by means of sensitivity analysis and the implementation of optimization strategies.
- Energy/exergy analysis uses the data achieved in the process simulation to provide a detailed analysis of energy and exergy losses for each unit composing the process, aimed to understand where there are the greatest energy/exergy penalties and how to reduce them.

- LCA is focused on the assessment of environmental footprint in the whole life cycle of the plant, from construction to operation and end-of-life. This impact is calculated by means of a series of indices, such as the Global Warming Potential, the Acidification Potential, the Eutrophication Potential and the Eco-toxicity.

In the following, a brief description of these process analysis approaches is reported.

7.2.1 Process simulation

To simulate a process, the following actions have to be made [9]:

1. Design specifications and battery limit selection.
2. Model Development: development of the process flow diagram (PFD), analysis of degrees of freedom, material and energy balances, selection of the thermodynamics model to be implemented.
3. Process Simulation: sensitivity analysis, optimization, basic design.
4. Post-processing activities: techno-economic analysis, dynamic simulations, exergetic analysis, LCA.

Figure 7.1 shows the procedure schematically.

The basis of a proper process modeling and simulation is the application of the material and energy balances. These equations derive from the mass and energy conservation principles and are formulated as follow, after the selection of the control volume where applying them:

Inlet rate of the quantity (mass or energy) through the control volume surface –
Outlet rate of the quantity (mass or energy) through the control volume surface ±
Generation rate of mass or energy inside the control volume =
Time variation of the quantity (mass or energy) in the control volume

Mathematical formulation can be expressed as follows:

Mass balance

$$\dot{m}_{in} - \dot{m}_{out} \pm \dot{m}_{gen} = \frac{dM}{dt} \tag{7.1}$$

where $\dot{m}_{in}$, $\dot{m}_{out}$ and $\dot{m}_{gen}$ are the inlet, outlet and generation flow rates, respectively, and M is the mass inside the control volume.

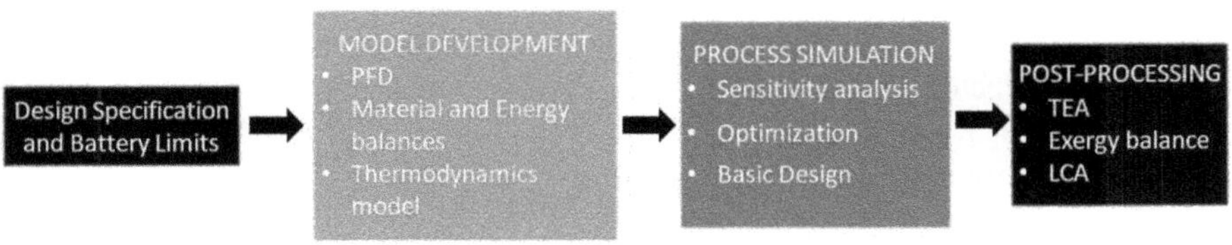

FIGURE 7.1 Process simulation procedure.

Energy balance

$$\dot{H}_{in} - \dot{H}_{out} \pm \dot{H}_{gen} = \frac{dH}{dt} \tag{7.2}$$

where $\dot{H}_{in}$, $\dot{H}_{out}$ and $\dot{H}_{gen}$ the inlet, outlet and generation enthalpy flow rates, respectively, and H is the enthalpy inside the control volume.

After formulating the mass and energy balances for each unit of the PFD, and for the whole process as well, an appropriate thermodynamic package has to be selected. There is not a single general thermodynamic model suitable in every situation, but rather selecting a model inappropriate for the specific application compromises the entire modeling. Main thermodynamics models are:

- Equations of State
- Activity Models
- Semi-Empirical Models
- Ad-hoc packages

The main Equations of State models and some of their application fields are:

- Peng-Robinson – Gas Dehydration, slightly polar liquids and light gases.
- PRSV – Cryogenic gas processing, vacuum towers.
- Soave Redlich Kwong – polar components at pressure <1 MPa.

Activity models are:

- Wilson and Margules – alcohols, phenols, C4-C6 hydrocarbons, aromatics.
- NRTL – aqueous organics.
- UNIQUAC and UNIFAC – polar components at pressure >1 MPa.

In addition, Semi-Empirical Models and Ad-hoc packages have been developed for specific industrial processes, such as sour gas absorption by means of amine solutions.

Then, after modeling activities, the process can be simulated through the use of simulation tools, such as Aspen Plus, PRO II, ASCEND and many others. First of all, the models should be validated with experimental data, when available. Then, after confirming the reliability of the models, we can perform a sensitivity analysis, by which it is possible to quantify the impact of the main operating conditions (temperature, pressure, composition, flowrates, etc.) on the process performance. The sensitivity analysis gives information for optimizing the operating conditions set, according to selected criteria such as:

- Maximization of process performance.
- Minimization of operating costs.
- Minimization of resource consumption.
- Minimization of environmental impact.

The designers should decide the optimization criteria and then select the optimal operating conditions set coherently with their needs. Then, post-processing activities are performed to verify the most important results achieved in terms of economic, environmental impact and efficiency of the process. In the following, two post-processing actions are described: the Exergy Analysis and the Life Cycle Assessment.

7.2.2 Exergy Analysis

The exergy of a system is the maximum obtainable work during a process that brings the system into equilibrium with its reference environment through a series of reversible processes in which the system can only interact with such environment [10–12]. In other words, exergy is an "opportunity for doing work" and, in real processes, it is destroyed by entropy generation [12]. Exergy-based analysis is useful to evaluate the thermodynamic inefficiencies of processes, to understand and locate the additional consumption of fuels or primary energy, to provide an instrument for comparison among different process configurations and to detect solution to reduce the energy penalties of a process.

This paragraph is focused on the fundamental definitions at the basis of exergy analysis thanks to the simultaneous invocation of the first and second laws.

Figure 7.2 represents a "black-box separator" (open system) in thermal contact with a number l of heat reservoirs (l = 1…p) at temperatures T_l. The mechanical power $\dot{W}$ is considered in this chapter, nevertheless this work transfer rate can be the combination of different effects (e.g. shear, electrical, magnetic) [14, 15]. The control volume is crossed by j feed streams and k product streams and the system is free to transfer work (p_0 dV/dt) and heat (Q_0) to the environment.

The First and Second Laws of Thermodynamics for the open system of Figure 7.2 result into Eqs. 7.3 and 7.4. The enthalpy is used by neglecting variation of kinetic and potential energy of the streams crossing the control volume.

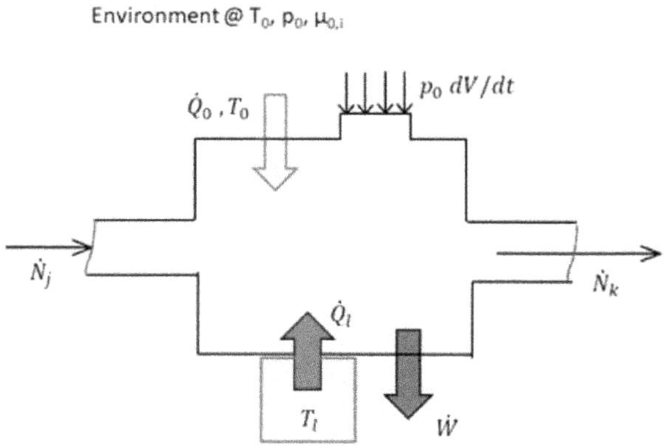

FIGURE 7.2 Control volume black-box separator adapted from Bejan [13].

$$\frac{dU}{dt} = \dot{W} + \sum_{l=0}^{p} \dot{Q}_l + \sum_{j=1}^{q} \dot{N}_j \bar{h}_j - \sum_{k=1}^{r} \dot{N}_k \bar{h}_k - p_0 \frac{dV}{dt} \tag{7.3}$$

$$\frac{dS}{dt} = \sum_{l=0}^{p} \frac{\dot{Q}_l}{T_l} + \sum_{j=1}^{q} \dot{N}_j \bar{s}_j - \sum_{k=1}^{r} \dot{N}_k \bar{s}_k + \dot{S}_{gen} \tag{7.4}$$

where N_j (with $j = 1 \ldots q$) and N_k (with $k = 1 \ldots r$) are the input and output streams, p_0 and T_0 are the environmental conditions, $\dot{S}_{gen}$ is the entropy generation. In the case of interest where the atmospheric work is absent, the equations are arranged to obtain the exergy balance for the open system of Eq. 7.5 where the specific exergy of any stream is given by Eq. 7.6.

$$\frac{d\Xi}{dt} = \dot{W} + \sum_{l=0}^{p} \left(1 - \frac{T_0}{T_l}\right)_l \dot{Q}_l + \sum_{j=1}^{q} \dot{N}_j \bar{\xi}_j - \sum_{k=1}^{r} \dot{N}_k \bar{\xi}_k - T_0 \dot{S}_{gen} \tag{7.5}$$

$$\xi_j(T, p, N_i) = h_j - T_o s_j - \sum_i \mu_{0,i} x_i \tag{7.6}$$

where for the j-th stream, ξ_j is the specific exergy, $\mu_{0,i}$ is chemical potential and x_i the molar fraction. In analogy, the terms for the output streams can be defined. The exergy of the system can be seen as the sum of the "available work" composed by the mechanical power and "adjusted" thermal power plus the specific exergy of the process streams multiplied by the stream flowrate. In fact, main interactions though the system boundaries are the work and heat-related exergy fluxes (respectively Ξ_W and Ξ_Q) while the exergy destruction term Ξ_D represents the irreversible loss of "work availability", equal to the so-called Gouy–Stodola lost work as reported in the following equation:

$$\Xi_W = \dot{W} \tag{7.7}$$

$$\Xi_Q = \left(1 - \frac{T_0}{T_l}\right) \tag{7.8}$$

$$\Xi_D = T_0 \dot{S}_{gen} \tag{7.9}$$

As often occurs for evaluating systems that includes separation and/or conversion processes, the balance of Eq. 7.5 neglects a portion of the physical exergy, related to the kinetic and potential energy for the sake of simplicity. A more rigorous definition including all terms as depicted in Figure 7.3 is given by Eq. 7.10

$$\xi(T, p, N_i, v, z) = \hat{h} - T_o s - \sum_i \mu_{0,i} x_i = gz + \frac{1}{2} v^2 + h - T_o s - \sum_i \mu_{0,i} x_i \tag{7.10}$$

where the additional first and the second terms are respectively the potential and kinetic energy forms of the stream.

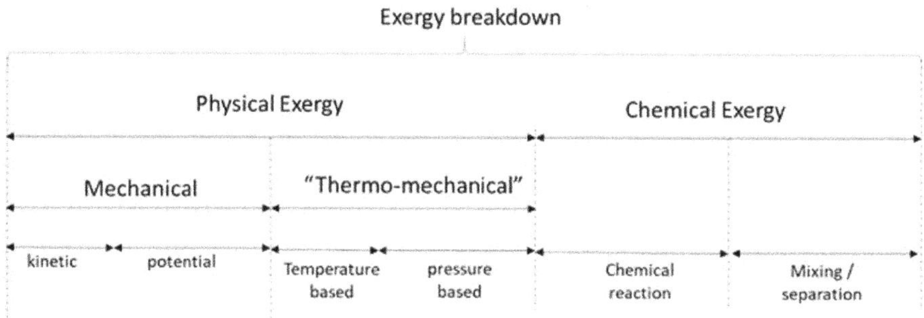

FIGURE 7.3 Exergy breakdown.

In synthesis, the evaluation of the entropy generation and the consequent destruction of the available work in any part of the process (by applying the exergy balance of Eq. 7.5) is the main goal of the exergy analysis and is implemented in techno-economic evaluations of separation processes in several papers [10–13, 16].

Before performing an exergy analysis it is important to [11, 12]:

- Univocally define the system by selecting the control volume.
- Define an appropriate reference environment.
- Compute mass, energy and entropy balances of the system during the process.
- In case of work-absorbing process (conversion-separation), evaluate the minimum necessary work.

The minimum work of separation can be evaluated from Eq. 7.5 by assuming steady state and reversible behavior and under the hypothesis that the process streams must exit at the "restricted dead state" [15]. In this way, the exergy of the process streams reduces to the Gibbs free energy at atmospheric temperature and pressure (Eq. 7.11).

$$\dot{W}_{least} = \dot{\Xi}_{least} = \sum_{k=1}^{r} \dot{N}_k \bar{g}_k - \sum_{j=1}^{q} \dot{N}_j \bar{g}_j \tag{7.11}$$

where $\dot{\Xi}_{least}$ is the minimum amount of exergy (rate of available energy) necessary to separate the streams N_j into the streams N_k and g is the specific Gibbs free energy of the j-th or k-th component. This minimum amount must be provided to the system by a combination of work and heat (for a stationary process that does not accumulate nonflow exergy, $d\Xi/dt = 0$). For example, with respect to the product "pure" stream obtained through an ideal separation box, the minimum work of separation is around 2 kJ/kg for seawater desalination, 480 kJ/kg to recovery the CO_2 from flue gases and 300 kJ/kg to separate the O_2 from air. More details can be found in the cited literature [9, 13, 15, 16].

In addition, to identifying the efficiencies, the exergy analysis, starting from the definitions above, allows the calculation of performance indices based on the II Law of Thermodynamics. Two similar definitions are reported below. The first, reported

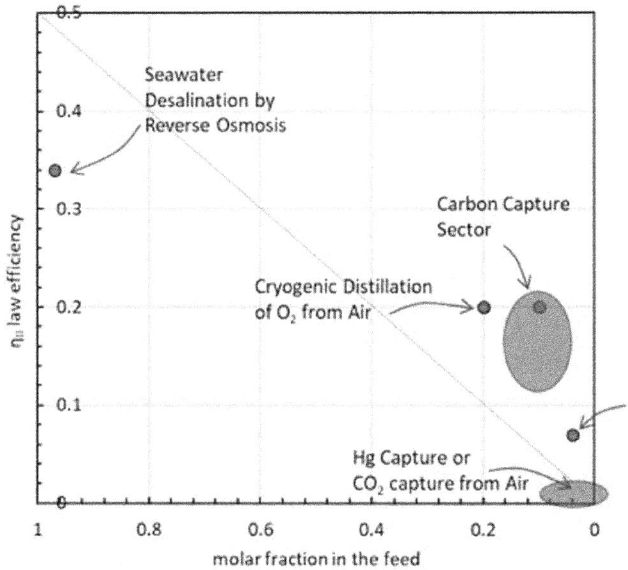

FIGURE 7.4 Second law efficiency vs. molar fractions in the feed to be separated.

in Eq. 7.12, is specifically designed for separation/conversion processes; the second is appliable to a general process and is defined according to the Fuel–Product paradigm (F–P) in which the fuel represents the primary energy input to the system and the useful exergy of the main product stream (assuming that the others are waste) [10–13]. Figure 7.4 shows the 2-nd law efficiency for reference processes as described by the cited literature [9, 15]. For mature processes, it is of the order of magnitude of 10–20% and it strongly decreases as the concentration of the substance to recover (or to abate in case of pollutants) decreases.

$$\eta_{II} \frac{W_{least}}{W_{real}} = \frac{Q_{least}}{Q_{real}} \tag{7.12}$$

$$\eta_{II} = \frac{\xi_{useful}}{\xi_{in}} = \frac{\xi_{product}}{\xi_{fuel}} = 1 - \frac{\xi_{des} + \xi_{lost}}{\xi_{fuel}} \tag{7.13}$$

7.3 LCA ANALYSIS

All products and services have some impact on the environment, different at each stage, determined by various aspects, depending on how the product was created, how long it will be used, how many resources and energy were used in process of its production and how will look final disposal of waste. Each stage has specific impact on the environment and efforts are currently being made to minimize this impact, especially where it is ecologically and economically significant. In this way, it is possible not only to influence the environmental factors positively, but also to decrease the costs of the production as well as services. Life Cycle Assessment is a reliable tool to evaluate the environmental aspects and impacts that result from the various stages of a product's life, including raw material acquisition (for example: mineral extraction and processing), manufacturing (production process), distribution and transport, usage, reuse, recycling, final disposal of waste.

Life Cycle Assessment can quantify the environmental impact on many different categories of products or services on every life cycle stage.

Because the LCA is a cradle-to-grave study, it allows to evaluate both the overall cycle's impact and the impact of individual process's stage or product. In addition, the LCA makes possible to understanding how entire process works on all stages, establishing complex and extensive relationships between various inputs, such as raw materials or energy and outputs which can be understand as intermediate products (necessary for further processes or technologies analysis) or as final products in their last phase of life, i.e., on disposal. LCA does not only indicate the environmental impact of a product cycle, but also on human health or on the depletion of natural resources. By studying the impacts of a production process or technology at each stage, the LCA can be used as a comparative tool for different scenarios, which can be applied in the decision support process [17].

Figure 7.5 represents a simple example of factors and relations between them that can be estimated in LCA analysis [18].

Fundamental tasks that can be managed with using LCA as a tool are: determine impact assessment on different factors like environment, human health, natural resources depleted, impact of wastes disposal, managing variant scenarios of processes, technologies and services, help in decision-making process and product or process design.

However, like any methodology, LCA analysis also has its limitations. The holistic approach is both a major strength as well as a limitation which may lead to a simplification of the analysis. It is important to define the purpose of the LCA and what simplifications can be applied.

Further limitations that should be taken into consideration for LCA analysis are: linear nature of the model, impossibility to quantify local environmental impacts (only global), limitation of the databases to particular processes in the form of

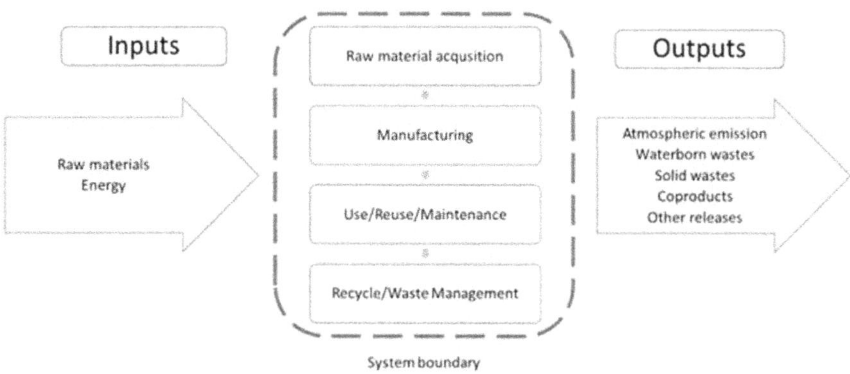

FIGURE 7.5 LCA analysis simplified scheme.

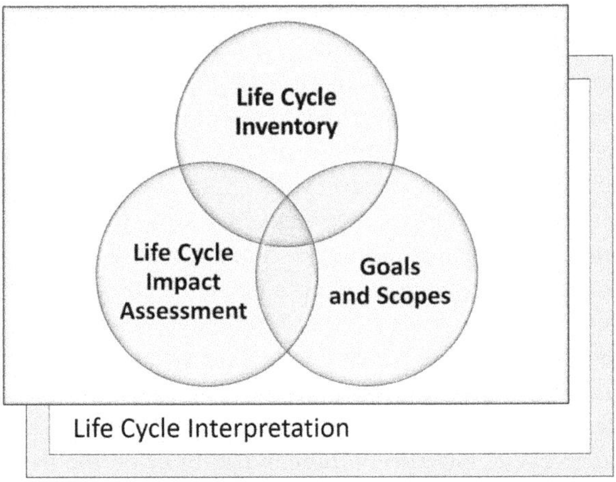

FIGURE 7.6 LCA steps.

predefined blocks, without the possibility to break them down into individual components (e.g. "electricity production from fossil fuels" as one element), limitation of access to detailed project data [19].

LCA analysis is composed of four major phases of process which are shown in Figure 7.6, including compilation and evaluation of the inputs, outputs and impact assessment with all elements related.

The first part of the LCA process is to define goals and scopes, which can provide data: how the research will be done, what is data assumed to achieved, identification of system boundary and data quality requirement as well as identification of the reasons for undertaking the study and the intended public.

The second phase is Life Cycle Inventory (LCI) which identifies data such as inputs and outputs. At this stage, all data should be collected, from the acquisition of raw materials and energy to the last wastes produced during the entire process. Although, the data obtained for this stage may vary due to pre-defined goals and scopes. Frequently, limited and very case study-specific data is sufficient for a particular instance.

Life Cycle Impact Assessment (LCIA) is the third phase of LCA process, strictly linked with LCI. It presented how particular elements from LCI have certain impacts.

There are three mandatory sub-stages of LCIA. The activities of the first stage are: selection of impact categories, category indicators and characterization models. The next step is a classification which assigned LCI elements to certain impact categories e.g. assigning CO_2 emission to global warming category. The third step is the characterization to calculate the contribution of classified LCI elements, within its impact category. To obtain this calculation, the value of each LCI element must be multiplied by the appropriate conversion factor for each considered impact category. The coefficients are specific to the substance or resource and represent the impact intensity of the element which is relative to the reference element, specific for the impact category (e.g. climate change will be estimated in CO_2 equivalent).

Then, normalization, which is an optional step, allows to compare obtained results to each other. This can be estimated by multiplying obtained previously results by normalization coefficients. Other optional steps are grouping and weighting, which both support understanding and interpretation of obtained data.

The last stage is not less important compared to previous steps. It is the Life Cycle Interpretation (LCI) and it combines the results from second and third process phases and compare them against goals and scopes from step one. It shows completeness, sensitivity and consistency checks as well as uncertainty and accuracy of obtained results. Additionally, it presents interpretation of environmental impacts [17, 20].

LCA is a crucial tool to quantify the real environmental benefits derived from the application of Carbon Capture Utilization and Storage (CCUS) technologies. The LCA analysis allows to evaluate the environmental impact of the entire processes, from the CO_2 separation unit to its compression and storage or its application in chemical processes to synthesize added-value products [21].

The decision to implement CCSU technology should be based on the knowledge provided by the results of in-depth analyses of the overall environmental impact and benefits of the technology, not just its effectiveness in capturing CO_2. LCA is an essential tool to provide this knowledge [22].

However, even if LCA analyses have been standardized in ISO 1404x, there is still considerable openness to a number of interpretations, e.g. in the choice of units, boundary conditions, methodology of environmental impact assessment and background processes. Many case study results for CO_2 capture and utilization and storage technologies cannot be compared, as all depends on the specification of the case, the complexity of the model developed and the applied environmental impact assessment methods. As a logical conclusion, such analysis should be undertaken on a project-specific issue. According to [19], many case studies in which similar boundary conditions and CCUS technologies were even assumed received different results precisely because of the difference in the choice of e.g. environmental assessment methodology or the choice of the type of input data. Nonetheless, LCA is intended to promote comparable and transparent research on the environmental impacts of CCSU.

7.4 CASE STUDY – PROCESS SIMULATION AND LCA OF THE DME SYNTHESIS FROM PURE CO_2 AND RENEWABLE ENERGY

Two process schemes are presented here and simulated by means of Aspen Plus. The two process configurations treat a flue gas containing carbon dioxide: the first takes as feed the flue gases coming from a power plant, the second treats the flue gases from a cement plant. In both the schemes, the carbon dioxide is first separated from the flue gas stream and then transformed in a high valuable product such as DME. Currently, the primary usage of DME is as an aerosol propellant in personal care products, such as hair sprays [23, 24]. At ambient condition this substance is a vapor, but it liquefies at 0.6 MPa and room temperature [25], making it very easy to store.

DME is likely to be utilized as a substitute for diesel fuel in diesel engines and gas turbines, as fuel for domestic usage or as intermediate in the production of hydrocarbon fuels and chemicals [23]. Through the reaction with sulfur trioxide it produces dimethyl sulfate [24]. It represents a key building block in the production of light olefins [26, 27], ethanol [28] and methyl acetate [29].

7.4.1 Power plant-based process

The process of conversion of carbon dioxide from the power plant to DME is composed of five macro units which are here listed:

1. CO_2 Separation and Purification.
2. Biomass Pyrolysis.
3. Renewable Energy and Green Hydrogen Production.
4. DME Production.
5. Reaction Products Separation.

These units are represented in Figures 7.7–7.12, where all components are reported and tagged in Aspen Plus environment.

7.4.1.1 CO_2 separation and purification

The flue gas from the power plant enters the DME production plant at 123°C with the composition reported in Table 1.1 in Chapter 1 [30].

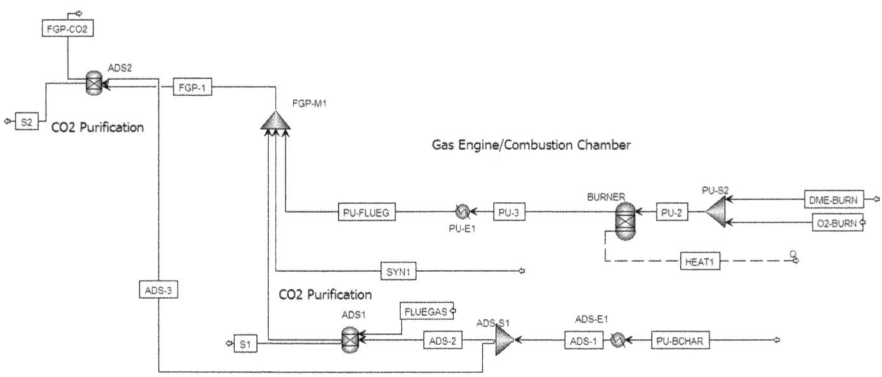

FIGURE 7.7 CO_2 separation and purification unit model in Aspen Plus (power plant).

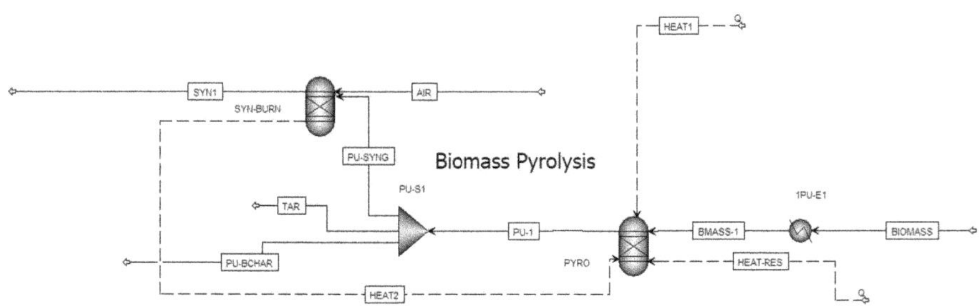

FIGURE 7.8 Biomass pyrolysis unit model in Aspen Plus.

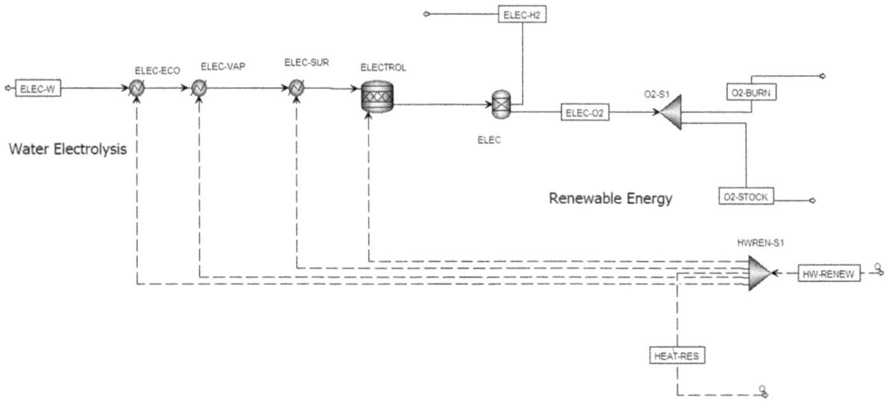

FIGURE 7.9 Green hydrogen production unit model in Aspen Plus.

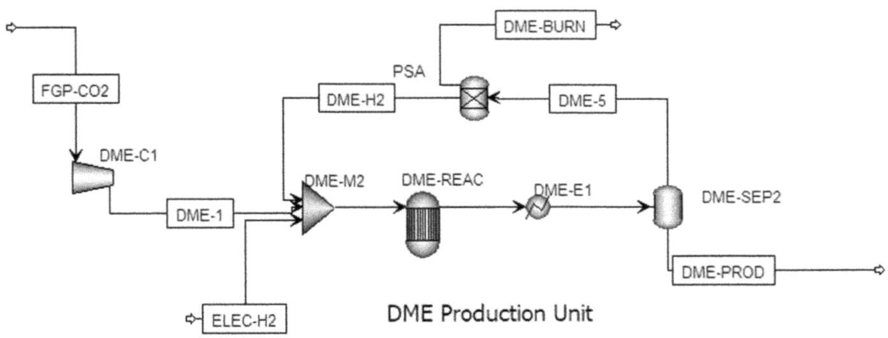

FIGURE 7.10 DME synthesis reactor model in Aspen Plus (power plant).

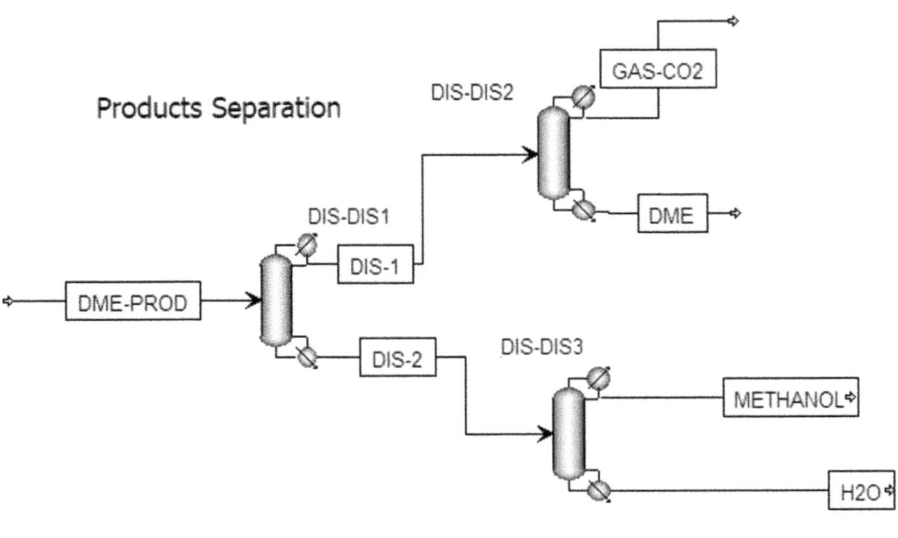

FIGURE 7.11 Products purification unit in Aspen Plus.

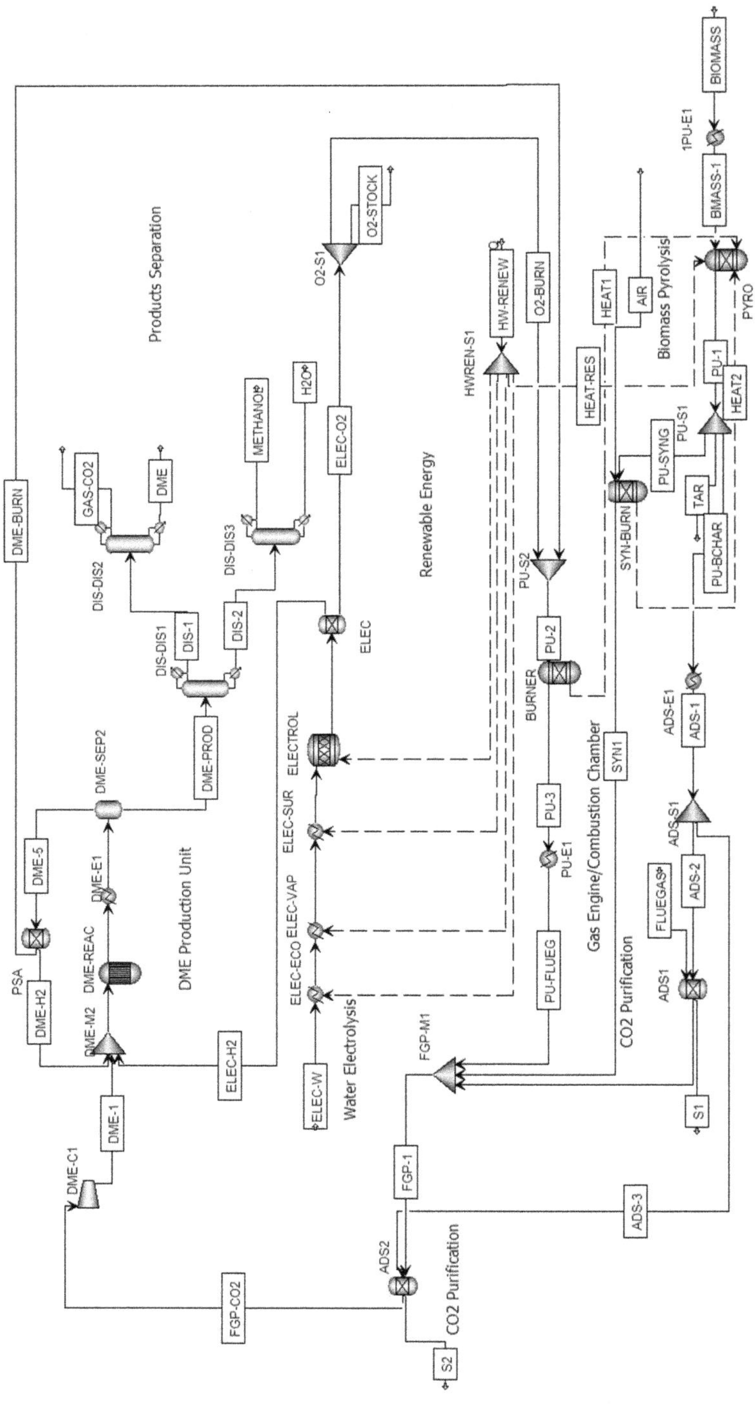

FIGURE 7.12 Whole process scheme in Aspen Plus (power plant).

Hence the amount of carbon dioxide in the flue gas is 7901 kmol/h (347672.41 kg/h) at 123°C and 130 kPa. The power plant was not simulated since the focus of the project is to use what usually is a waste (carbon dioxide) to produce a valuable product, such as DME.

The flue gas is treated to concentrate the CO_2. Another stream containing carbon dioxide is subject to purification as shown in Figure 7.7. This stream consists of the oxyfuel combustion products of the un-reacted gas exiting the DME reactor and contains 3957.57 kmol/h of CO_2. As a consequence, the carbon dioxide fed to the reactor is nearly doubled.

The flue gas purification and carbon dioxide adsorption consist of a series of adsorbent columns in which the flue gas is first purified, then carbon dioxide is separated from the other gases and lastly purified by further adsorption. The adsorbent comes from the pyrolysis of the biomass. The process of CO_2 isolating and purifying is simulated in Aspen Plus environment with the Aspen Plus block Sep. Sep separates components based on specified flows or split fractions.

7.4.1.2 Biomass pyrolysis

The waste biomass consists of wood, molecular formula $CH_{1.46}O_{0.689}$ [30]. First, the biomass is dried up, as shown in Figure 7.8; the heat is taken from the hot gas exiting the combustion chamber in the DME Production Unit. Then the biomass enters the pyrolysis reactor where the reaction heat is supplied by the hot gas from the combustion chamber in the DME Production Unit and by the combustion of the syngas, obtained as a byproduct of the pyrolysis process itself (the heat supplied by the combustion is not sufficient to thermally treat the biomass). The combusted syngas, as well as the hot gas from the combustion chamber, is fed to the Flue Gas Purification Unit to separate the CO_2 contained in it. The biochar leaves the pyrolysis reactor at 800°C [30] and undergoes an activation process; the adsorbent is used in the purification of the flue gas and CO_2 adsorption. The pyrolysis is not simulated in Aspen Plus, since data are taken from literature.

The purification is performed by adsorption with the biochar obtained in the waste biomass pyrolysis unit; the biochar has the following molecular formula $CH_{0.153}O_{0.0324}$ [30]. The adsorbent is able to isolate 80% of the carbon dioxide in the remaining flue gas 80%, providing a stream of CO_2 at 99% molar purity (the remaining is molecular oxygen). This is simulated in Aspen Plus with the block Sep. The flowrate and purity of the carbon dioxide leaving the block Sep are set according to the values above. The purified CO_2 enters the unit of DME production where it is converted into DME and MeOH (methanol, as a byproduct) through reaction with H_2.

7.4.1.3 Renewable energy and green hydrogen production

The hydrogen is provided by water electrolysis powered by RES (Renewable Energy Sources), making it the so called "green hydrogen". For a PEMEC (Proton Exchange Membrane Electrolyzer), the power consumption to produce H_2 at 3 MPa is around a value of 4.5 kWh/Nm^3 H_2 [31] with a purity of 99.999% for both H_2 and O_2 [32]. The water electrolysis is not simulated in Aspen Plus, a stream of H_2 and a stream of O_2 are added at the flowsheet with a molar ratio of 2:1 (H_2/O_2) and an electric power demand is taken into account. The un-reacted gas exiting the DME reactor is previously sent to a PSA that recovers 86% of the hydrogen in it. The hydrogen recovered at the PSA is fed to the DME reactor; this allows to reduce the quantity of hydrogen produced by the electrolyzer at 12473.12 kg/h. Hence the power consumption at the electrolyzer is 629.04 MW. The PSA unit is simulated with a Sep in Aspen Plus.

The oxygen is consumed in the oxyfuel combustion of the un-reacted gas stream exiting the DME reactor. The combustion of the un-reacted gas provides further carbon dioxide which is previously separated from the other gases and then fed to the DME reactor. Possible overproduction of hydrogen and oxygen is stored.

7.4.1.4 DME production

The carbon dioxide feeding the DME reactor is compressed to 3 MPa and then it is mixed with the hydrogen produced via water electrolysis to obtain a H_2/CO_2 ratio of 3 [33] (see Figure 7.10). The resulting gas stream is heated up to 200°C and sent to the DME reactor. CO_2 is partially converted into DME and MeOH in a single-step reactor. The reactor was simulated in Aspen Plus with a RGibbs reactor and with REquil reactor. The former provides rigorous reaction and multiphase equilibrium based on Gibbs free energy minimization. The latter allows rigorous equilibrium reactor based on stoichiometric approach. The results of those reactors show absolute symmetry; thus the RGibbs reactor was chosen as easier to manage.

The reaction products are cooled down to 30°C, thus allowing the separation of the desired products, DME and MeOH, from the un-reacted gas via liquid-gas separation. The liquid is an aqueous mixture whose molar composition is 6.25% of DME and 7.31% of MeOH, the gas is a mixture of the un-reacted gases. The gas stream, containing the unreacted gases, is fed to a PSA (pressure swing adsorption unit) to recycle the H_2 to the reactor. The remaining gases perform an oxyfuel combustion with O_2 produced via water electrolysis, the combustion chamber is simulated with a RGibbs reactor. The liquid products are sent to the Reaction Products Separation Unit.

7.4.1.5 Reaction products separation

As shown in Figure 7.11, the unit is essentially made of three distillation columns. Each of the three distillation columns was simulated in Aspen Plus with the block RadFrac, which provides rigorous 2 or 3-phase fractionation for a single column.

The liquid mixture containing DME and MeOH is sent to the first distillation column operating at 2 MPa. This column performs the separation of DME from MeOH. The former exits from the head of the column in a stream containing the other gases (mostly unreacted carbon dioxide), while the latter exits from the bottom of the column in an aqueous mixture.

The distillate of the first distillation column is sent to the second one operating at 1.9 MPa. The distillate of the second column is composed of unreacted CO_2 mostly. DME is collected at the bottom of the column with a flowrate of 203.16 kmol/h (9359.14 kg/h) and a molar purity of 99.98% (mass purity is 99.99%). The impurities in the DME stream are:

- 0.02% MeOH molar;
- ppm CO_2.

The bottoms of the first distillation column are sent to a third one operating at a much lower pressure (200 kPa). Here, the aqueous mixture containing MeOH is separated: MeOH is obtained as the distillate product, and the bottoms of the column collect the water with a 0.07 mol% of MeOH (0.13 wt%). The flowrate of the distillate of this column is 173.83 kmol/h (5547 kg/h) and contains MeOH with a molar concentration of 99.06% (mass purity is 99.47%), the remaining share is water (0.94 mol%).

7.4.2 Cement plant-based process scheme

In this process, the DME is produced from the carbon dioxide contained in the kiln gas of a cement plant. The scheme process is organized into five units listed below:

1. CO_2 Separation and Purification and Concrete Curing Carbonation.
2. Biomass Pyrolysis.
3. Renewable Energy and Green Hydrogen Production.
4. DME Production.
5. Reaction Products Separation.

The main difference between the power plant process scheme and the cement plant process scheme stands in the CO_2 Separation and Purification Unit. However, the components of the macro units are the same, as shown in Figures 7.8, 7.9, 7.11, 7.13 and 7.14. Figure 7.15 shows the whole process scheme modeled in Aspen environment.

7.4.2.1 CO_2 separation and purification and concrete curing carbonation

The cement plant was not simulated in Aspen Plus for the same reasons the power plant was not too. The flue gas from the cement has the composition and flowrate reported in Table 1.2 in Chapter 1 [35].

Similarly to what happens in the scheme of the power plant, the flue gas is treated in order to be purified and to concentrate the CO_2 in it. The same is done to the combusted gas generated by the oxyfuel combustion of the un-reacted gas fed to the DME reactor. The flue gas purification and CO_2 separation are performed with the same series of adsorbing step of the power plant scheme; the adsorbent is obtained by the biomass pyrolysis. In this process, a fraction of carbon dioxide is used to perform concrete curing carbonation, an accelerated curing process that injects CO_2 gas into the curing vessel and transforms the gaseous CO_2 into solid calcium carbonates [34]. The process of concrete curing carbonation is not simulated; in Aspen Plus, a fraction of the purified CO_2 obtained by adsorption, representing the feed to the curing processes, is split from the stream that enters the DME Production Unit.

The process of CO_2 purification is simulated in Aspen Plus with the block Sep.

7.4.2.2 Biomass pyrolysis

The waste biomass used is the same of the power plant scheme and undergoes the same processes. First, it is dried up using the hot gas from the combustion chamber in the DME Production Unit. Then the biomass enters the pyrolysis reactor and the reaction heat is supplied by the hot gas from the combustion chamber in the

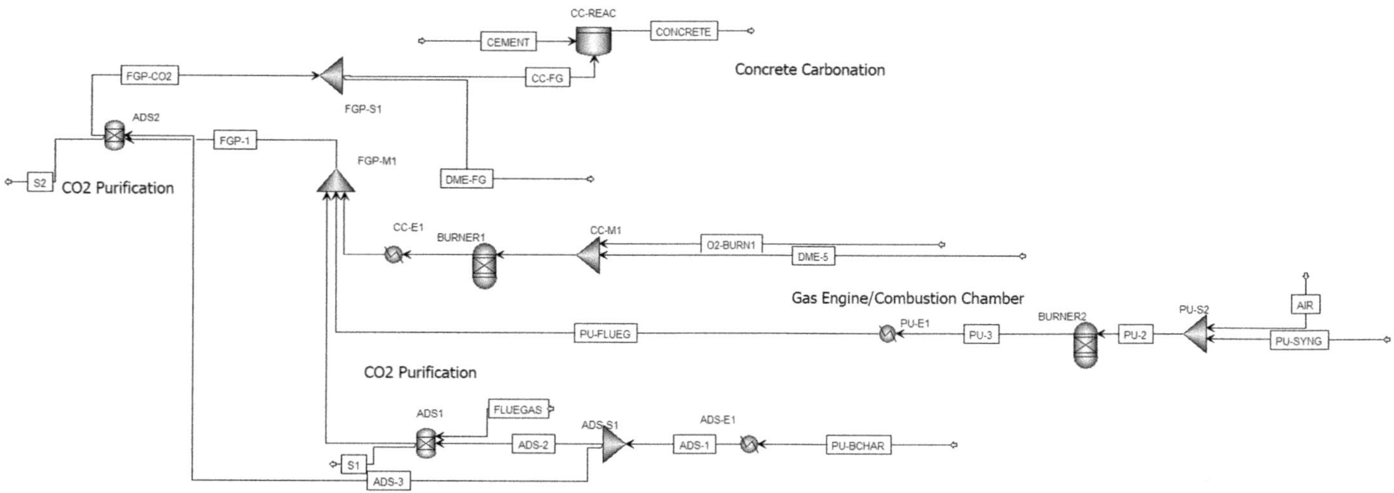

FIGURE 7.13 CO_2 separation and purification and Concrete Curing Carbonation unit model in Aspen Plus (cement plant).

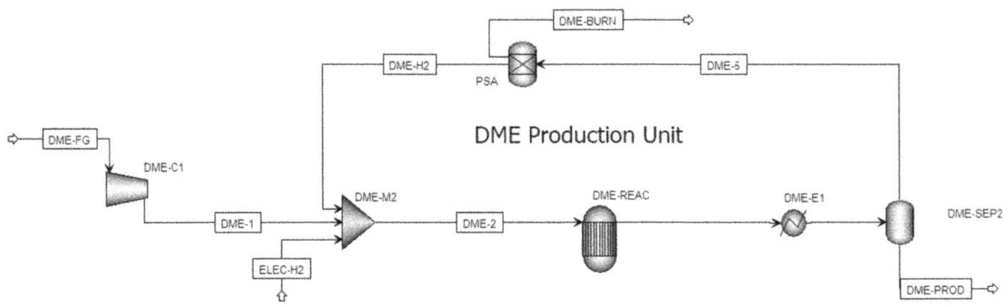

FIGURE 7.14 DME synthesis reactor model in Aspen Plus (cement plant).

DME Production Unit and by the combustion of the syngas obtained as a byproduct of the pyrolysis process itself, by the way the heat generated is not enough to perform the complete thermal treatment. Those hot gases are both sent to the Flue Gas Purification Unit to isolate the CO_2 contained in them. The biochar exits the pyrolysis reactor at 800°C [30] and is activated to be ready to adsorb the carbon dioxide. The pyrolysis is not simulated; data are taken from the literature.

The activated adsorbent is able to extract the 80% of the CO_2 in the gas stream with a 99% of molar purity (the remaining share in molar oxygen), this is simulated in Aspen Plus with the block Sep. The purified CO_2 enters the unit of DME production where it is converted into DME and MeOH through the reaction with green hydrogen.

7.4.2.3 Renewable energy and green hydrogen production

This unit is the same for the Power Plant-based process and for the Cement Plant-based process, thus it is not described here. The power consumption due to the water electrolysis is 717.92 MW.

7.4.2.4 DME production

CO_2, coming from the series of adsorbing steps, hence purified, is compressed to 3 MPa and mixed with the green-H_2 from water electrolysis in a molar ratio H_2/CO_2 of 3. This gaseous mixture is heated up to 200°C and fed to the DME reactor. Due to the reaction, CO_2 is partially converted into DME and MeOH in a single step-reactor simulated in Aspen Plus with a RGibbs reactor. The products exiting the reactor are cooled down to 30°C.

The reaction products are cooled down to 30°C; this allows the separation of the desired products, DME and MeOH, from the un-reacted gas via liquid-gas separation. The liquid is an aqueous mixture with a molar content of DME and MeOH of 7.31% and 6.25%, respectively, the gas is a mixture of the un-reacted gases. The gas stream, containing the unreacted gases, is sent to a PSA (pressure swing adsorption unit) to recycle the H_2 to the reactor, simulated with a Sep in Aspen Plus. The remaining gases perform an oxyfuel combustion with O_2 from the water electrolysis, combustion chamber is simulated with a RGibbs reactor. The liquid products are sent to the Reaction Products Separation Unit.

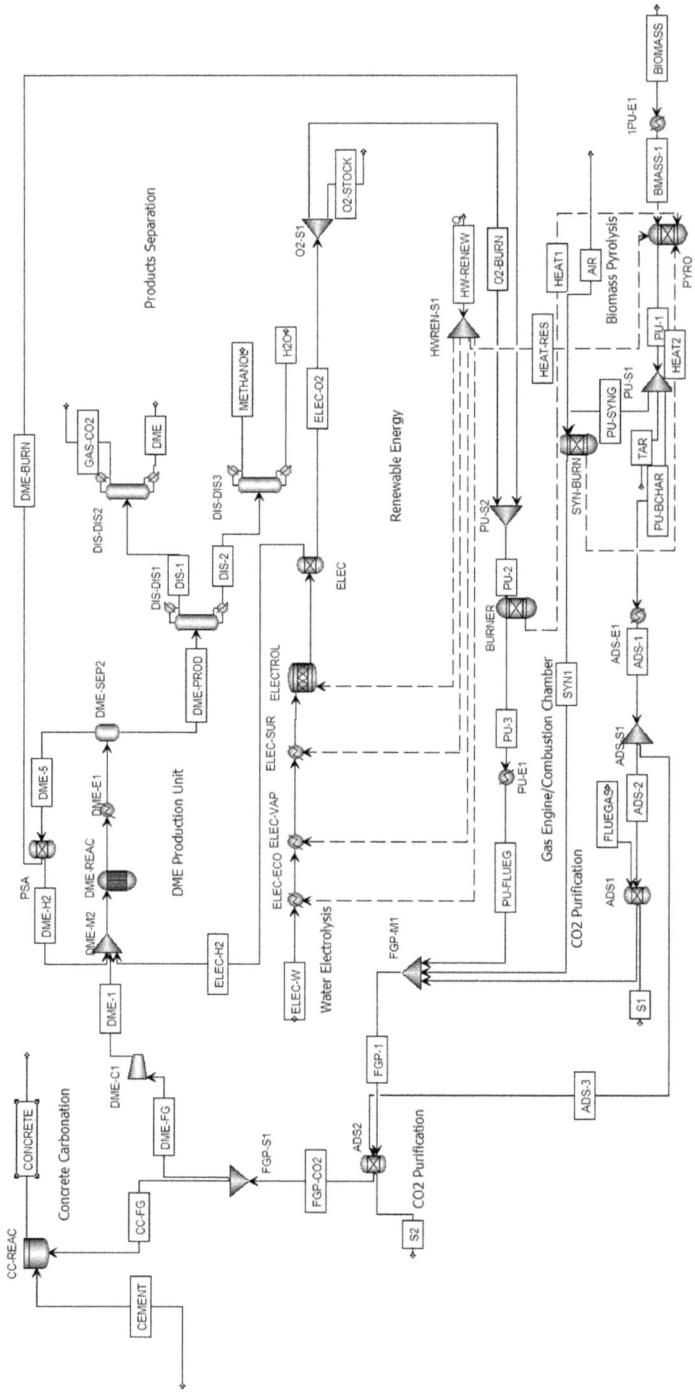

FIGURE 7.15 Whole process scheme (cement plant).

7.4.2.5 Reaction products separation

This unit is the same in both the Power Plant based process and the Cement Plant based process. Here just the results are listed, the description is not presented. The stream collected at the bottom of the second column has a flowrate of 231.75 kmol/h (10674 kg/h) and carries DME with a molar purity of 99.91% (mass purity is 99.94%). The impurities in the DME stream are due to MeOH and CO_2 presence (0.08% and 0.01 mol%, respectively).

The distillate of the third column has a flowrate of 197.96 kmol/h (6316.79 kg/h) and carries MeOH with a molar purity of 99.06% (mass purity is 99.47%). The remaining share is water (0.94 mol%). The stream collected at the bottom of the third column carries dirty water with a flowrate of 2526.88 kmol/h (45548.5 kg/h) in which the molar content of the water is 99.93% (99.87 wt%); the remaining share is MeOH.

Tables 7.1 and 7.2 summarize the main results of the process simulation of the two schemes.

7.5 LCA ANALYSIS OF THE PROCESS SCHEMES

A LCA analysis of the process schemes illustrated in Sections 7.4.1 and 7.4.2 are here reported following the methodology described in Section 7.3. The step of normalization and weighting score was not considered in this study. OpenLCA software by Green Delta has been applied for this simulation. Implemented databases are free to download as comes as follows:

- Ecoinvent 3.7 LCIA methods – EULA license – use of the free part of the database.
- ELCD 3.2 – free access.
- OpenLCA LCIA methods v.1.5.7 – Creative Commons Attribution – ShareAlike 4.0 International License.

The main objective of the LCA analysis is to evaluate the environmental impact of DME production with use of CO_2 captured from flue gases from power plant and cement plant. The scope of this study is to present a life cycle inventory of DME production, in order to allocate the environmental impact to each stage of its production (power plant/cement plant, capture and purification station, DME and/or carbonate concrete production). Furthermore, in the case of cement plant, the separation of flue gas stream for production of DME and for carbonate concrete is provided. The flue gas stream was split into two individual utilization methods according to Figure 1.4; however, for the purpose of calculating the Impact Assessment, the cement plant model is treated as a single object with implemented allocation for DME/carbonate concrete production.

The system boundaries were set based on Figures 1.3 and 1.4. Additionally, to simplify calculations, we made few cut outs and made a streamlined flowchart of the system, which is shown in Figure 7.16.

TABLE 7.1
Selected parameters of case study process simulation for power plant

	Mass flow	Composition								Renewable energy in	Heat energy in	Heat energy out
		CO_2	O_2	N_2	H_2	CO	DME	CH_3OH	H_2O			
Symbol	**kg/h**	**% mol**	**% mol**	**% mol**	**% mol**	**% mol**	**% mol**	**% mol**	**% mol**	**kW**	**MJ/h**	**MJ/h**
PU-flueg	248343	51.43	1.04		1.16	9.38			36.99	–	–	913179
Fluegas	1645200	14.30	7.50	68.30	Ar 0.80		–	–	9.10	–	–	–
FGP-CO_2/DME-1	239362	1	–	–	–	–	–	–	–	–	–	–
DME-H_2	20418.2		–	–	100	–	–	–	–	–	–	–
Elec-H_2	12473.1											–
DME-prod	62961.3	6.58			0.11	0.03	7.31	6.25	79.72			–
DME-5	209282	22.56			73.27	1.36	2.54	0.10	0.17	–	–	–
DME-burn	188864	60.98			27.73	3.68	6.86	0.29	0.46	–	–	–
Elec-W	112257.9	–	–	–	–	–	–	–	–	629037	–	–
Elec-O_2	99784.8	–	–	–	–	–	–	–	–	–	–	–
O_2-burn	59478.8	–	–	–	–	–	–	–	–	–	–	–
O_2-stock	40306	–	–	–	–	–	–	–	–	–	–	–
DME	9359.14	–	–	–	–	–	99.98	0.02	–			–
Methanol	5547	–	–	–	–	–	–	99.06	0.94			–
H_2O	39965.9	–	–	–	–	–	–	0.07	99.93			–
Gas-CO_2	8089.21	97.93	–	–	1.59	0.45	0.03	–	–			–

TABLE 7.2
Selected parameters of case study process simulation for cement plant with concrete carbonation

Symbol	Mass flow	Composition CO_2	O_2	N_2	H_2	CO	DME	CH_3OH	H_2O	Renewable energy in	Heat energy in	Heat energy out
	kg/h	% mol	% mol	% mol	% mol	% mol	% mol	% mol	% mol	kW	MJ/h	MJ/h
PU-Flueg	283144	51.43	1.04		1.16	9.38			36.99	–	–	1041191
Fluegas	432000	17.80	7.50	56.50	–	–	–	–	18.20	–	–	–
FGP-CO_2	286142	100	–	–	–	–	–	–	–	–	–	–
DME-FG	272805	100	–	–	–	–	–	–	–	–	–	–
DME-H_2	23279.3									–	–	–
Elec-H_2	14235.6									–	–	–
DME-prod	71783.9	6.58			0.11	0.03	7.31	6.25	79.72			–
DME-5	238608	22.56			73.27	1.36	2.54	0.10	0.17	–	–	–
DME-burn	215329	60.98			27.73	3.68	6.86	0.29	0.46		–	–
Elec-W	128120	–	–	–	–	–	–	–	–	717920	–	–
Elec-O_2	113884	–		–	–	–	–	–	–		–	–
O_2-burn	67815	–		–	–	–	–	–	–	–	–	–
O_2-stock	46069	–		–	–	–	–	–	–	–	–	–
DME	10674	0.01	–	–	–	–	99.91	0.08	–			–
Methanol	6316.79							99.06	0.94			–
H_2O	45548.5							0.07	99.93			–
Gas-CO_2	9216.51	97.95			1.59	0.45	0.01					–
CC-FG	13337	100	–	–	–	–	–	–	–	–	–	–
Cement	133370											
Concrete	666850											

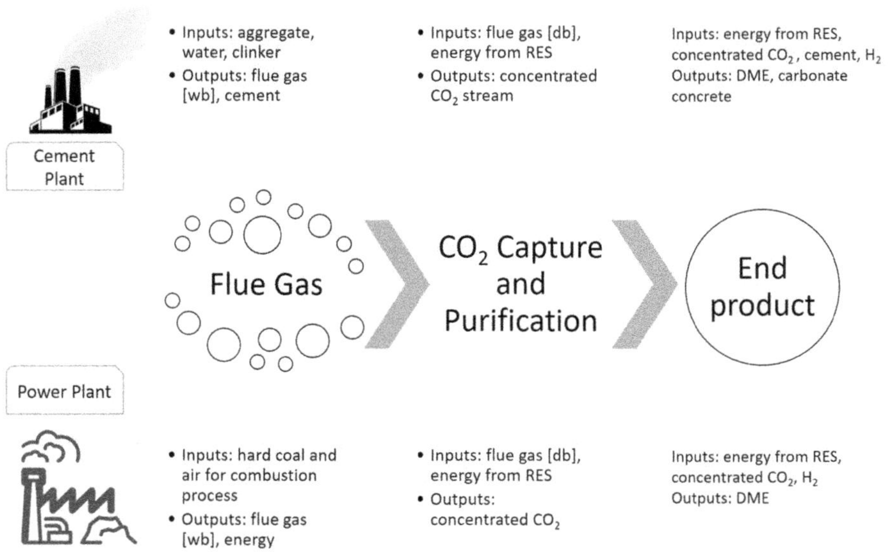

FIGURE 7.16 Life Cycle Assessment process.

All data necessary to perform the calculations have been collected in this book and refer to the tables: Tables 1.4, 3.4, 5.1 and 6.4. In this case study, all laboratory inputs have been scaled appropriately to industrial scale – for 1 hour of systems operation. The limitations and system boundaries of the model also affected the calculation of the Impact Assessment. The data collected are reported in Table 7.3.

The environmental impact from life cycle inventory is calculated using The International Reference Life Cycle Data System (ILCD) method (ILCD 2011, midpoint [v1.0.10, August 2016]) which determines 16 impacts such as climate change, ozone depletion, human toxicity (non-cancer and cancer effects), particulate matter (PM), ionizing radiation (HH), ionizing radiation E (interim), photochemical ozone formation (POF), acidification, terrestrial eutrophication, freshwater ecotoxicity, land use, water resource depletion, and mineral, fossil and resource depletion [36]. For our case study, we focus and discuss only on few of them as shown in Table 7.4.

The processes comparison is related to 1 hour of operation of the modeled facilities. It has to be noticed that the hourly DME production for the cement plant is almost 25% higher than for the power plant. The difference in the amount of DME produced depends on higher CO_2 flow rate from purification unit in case of cement plant. Although we obtain lower hourly DME production in the power plant model case, the impact in terms of carbon dioxide is lower (23.77 kg CO_2 eq. per 1 kg DME against 138.54 kg CO_2 eq. in the case of the cement plant). However, this is not a sufficiently accurate indicator because of the allocation of the stream for DME and carbonate concrete production, after the CO_2 capture and purification unit.

The overall environmental impact of DME production is higher for cement plant than for power plant. This is affected not only by the much larger amount of individual substances making up the outlet stream from DME production, but also due to considerably higher quantities of inputs in the outlet gas stream from the CO_2 capture unit (as shown in Table 7.3).

Even if process of carbon capture and utilization from power plant is not zero-emission it is still possible to reduce environmental impact of unit by implementation of CCU technology – and thus obtaining CO_2 as a resource in the production for green DME or carbonated concrete or other green technologies.

TABLE 7.3
Life Cycle Inventory – data were obtained from previous chapters and recalculated for industrial scale (in case for modeling purposes)

	Power plant	Cement plant
	Output from power plant [wb]	**Output from cement plant [wb]**
Parameter	kg/h	kg/h
Flue gas flow	457	120
CO_2	347672.41	115364.98
O_2	132615.03	35351.88
N_2	1056720.79	233027.82
H_2O	90509.76	48255.32
Ar	17682.00	n/a
Total	1645200.00	432000.00
	Recirculation from oxyfuel combustion	
Flue gas flow	62.98	90.08
CO_2	174133.29	198535.77
O_2	2561.15	2920.87
H_2O	51260.08	58443.15
H_2	180.12	205.35
CO	20208.20	23038.95
Total	248342.84	283144.08
	CO_2 capture and purification	
	Inlet from power plant [db]	**Inlet from cement plant [db]**
Energy demand [kWh]	545286.96	291927.69
Flue gas flow	486.60	178.08
CO_2	521805.70	313900.74
O_2	135176.18	38272.75
N_2	1056720.79	233027.82
Ar	17682.00	0.00
H_2	180.12	205.35
CO	20208.20	23038.95
Total	1751772.99	608445.61

(Continued)

TABLE 7.3 *(Continued)*

Life Cycle Inventory – data were obtained from previous chapters and recalculated for industrial scale (in case for modeling purposes)

	Outlet [db]	Outlet [db]	
Flue gas flow	72.94	87.95	
CO_2	239362.39	286214.07	
O_2	9361.37	1273.12	
N_2	13559.06	2265.10	
Ar	298.77	212.19	
Total	262581.58	289964.48	
	Production unit		
	DME	**DME**	**Concrete**
Production ratio		95.73%	4.27%
Energy demand [kWh]	51202.09	639.99	–
CO_2	239362.39	299175.53	13336.99
H_2 (total)	32891.28	41111.52	–
Electrolyzed	12473.12	15587.70	–
Recirculated from PSA	20418.16	25523.83	–
Total	272253.67	340287.05	13336.99
	DME output	**DME/carbonate concrete output**	
DME	28161.15	35199.64	–
MeOH	6125.54	7656.54	–
CO_2	167624.72	209520.23	–
CO	6155.49	7693.97	–
H_2	23748.04	29683.54	–
H_2O	40428.56	50533.13	–
Total	272243.50	340287.05	680186.56
Cement production		–	133369.91
Concrete production		–	666849.57

TABLE 7.4

Life Cycle Assessment results

Indicator	Power plant	Cement plant	Unit
Climate change	6.69476×10^5	4.87649×10^6	kg CO_2 eq.
Human toxicity – carcinogenics	3.06559×10^{-4}	7.69674×10^{-3}	CTUh
Human toxicity – non-carcinogenics	1.14970×10^{-2}	1.92157×10^{-1}	CTUh
Ozone depletion	7.32560×10^{-3}	2.79747×10^{-1}	kg CFC-11 eq.
Resource depletion – mineral, fossils and renewables	1.64542×10^3	2.31399×10^3	kg Sb eq.
Resource depletion – water	7.49116×10	8.42254×10^2	m^3

NOMENCLATURE

Symbols

g	Specific Gibbs free energy [kJ/kg]
$\bar{g}$	Molar Gibbs free energy [kJ/kmol]
h	Specific enthalpy [kJ/kg]
$\dot{H}$	Enthalpy flow rate [kJ/s]
$\bar{h}$	Molar enthalpy [kJ/kmol]
$\dot{m}$	Mass flow rate [kg/s]
N	Molar flow rate [kmol/s]
p	Pressure [kPa]
$\dot{Q}$	Heat flux [kW]
s	Specific entropy [kJ/kg K]
S_{gen}	Specific entropy generation per unit water produced [kJ/kg K]
$\dot{S}_{gen}$	Entropy generation rate [kW/K]
T	Temperature [°C]
$\dot{W}$	Work [kW]

Greek

ηpp	Power plant intrinsic efficiency [–]
ηII	Second Law/exergetic efficiency [–]
μ	Chemical potential [kJ/mol]
ξ	Specific exergy [kJ/kg]
$\dot{\Xi}$	Exergy flow rate [kW]

Subscripts & acronyms

0	Ambient
gen	Generated
in	Inlet
min	Minimum
out	Outlet

REFERENCES

[1] Centi, G.; De Falco, M.; Iaquaniello, G.; Perathoner, S. Strategy and drivers for CO_2 (re)use. In *CO_2: A valuable source of carbon, green energy and technology*; De Falco, M., Iaquaniello, G., Centi, G. Eds. Springer-Verlag, 2013, 1–26.

[2] Guo, J.; Lou, H.; Zhao, H.; Chai, D.; Zheng, X. Dry reforming of methane over nickel catalysts supported on magnesium aluminate spinels, *Applied Catalysis A: General* **2004**, *273, 75–82*, https://doi.org/10.1016/j.apcata.2004.06.014.

[3] Xin-Mei, L.; Lu, G. Q.; Zi-Feng, Y.; Beltramini, J. Recent advances in catalysts for methanol synthesis via hydrogenation of CO and CO_2, *Industrial & Engineering Chemistry Research* **2003**, *42*, *6518–6530*, https://doi.org/10.1021/ie020979s.

[4] Haiduc, A. G.; Brandenberger, M.; Suquet, S.; Vogel, F.; Bernier-Latmani, R.; Ludwig, C. SunCHem: An integrated process for the hydrothermal production of methane from microalgae and CO_2 mitigation, *Journal of Applied Psychology* **2009**, *21, 529–541*, https://doi.org/10.1007/s10811-009-9403-3.

[5] Barbarossa, V.; Bassano, C.; Deiana, P.; Vanga, G. CO_2 Conversion to CH_4. In *CO_2: A valuable source of carbon, green energy and technology*; De Falco, M., Iaquaniello, G., Centi, G., Eds. Springer-Verlag, 2013, 123–146.
[6] Mennicken, L.; Janz, A.; Roth, S. The German R&D program for CO_2 utilization—Innovations for a green economy, *Environmental Science and Pollution Research* **2016**, *23, 11386–11392*, https://doi.org/10.1007/s11356-016-6641-1.
[7] Arcoumanis, C.; Bae, C.; Crookes, R.; Kinoshita, E. The potential of di-methyl ether (DME) as an alternative fuel for compression-ignition engines: A review, *Fuel* **2008**, *87, 1014–1030*, https://doi.org/10.1016/j.fuel.2007.06.007.
[8] Ying, W.; Longbao, Z.; Hewu, W. Diesel emission improvements by the use of oxygenated DME/diesel blend fuels, *Atmospheric Environment* **2006**, *40, 2313–2320*, https://doi.org/10.1016/j.atmosenv.2005.12.016.
[9] De Falco M.; Capocelli, M. Process analysis and plant simulation in a sustainable economy, *Studies in Surface Science and Catalysis* **2019**, *179, 121–140*, https://doi.org/10.1016/B978-0-444-64337-7.00008-2.
[10] Moran, M. J.; Shapiro, H. N.; Boettner, D. D.; Bailey, M. *Fundamentals of engineering thermodynamics*. John Wiley & Sons; 2010.
[11] Kotas, T. J. *The exergy method of thermal plant analysis*. Paragon Publishing, 2012.
[12] Rocco, M. V.; Colombo, E.; Sciubba, E. Advances in exergy analysis: A novel assessment of the Extended Exergy Accounting method, *Applied Energy* **2014**, *113, 1405–1420*, https://doi.org/10.1016/j.apenergy.2013.08.080.
[13] Bejan, A. *Advanced engineering thermodynamics*, 4th ed. John Wiley & Sons, Inc., 2016.
[14] Olaleye, A. K.; Wang, M. Conventional and advanced exergy analysis of post-combustion CO_2 capture based on chemical absorption integrated with supercritical coal-fired power plant, *International Journal of Greenhouse Gas Control* **2017**, *64, 246–256*. ISSN 1750-5836, https://doi.org/10.1016/j.ijggc.2017.08.002.
[15] Wilcox, J. *Carbon capture*. Springer, 2012, ISBN 978-1-4614-2214-3.
[16] Mistry, K. H.; Lienhard, J. H. V. Generalized least energy of separation for desalination and other chemical separation processes, *Entropy* **2013**, *15, 2046–2080*, https://doi.org/10.3390/e15062046.
[17] Deska, M.; Kruczek, M.; Zawartka, P.; Burchart, D. Zastosowanie metody LCA do wspomagania decyzji środowiskowych w gospodarce wodno-ściekowej, *Scientific Papers of Silesian University of Technology. Organization and Management Series* **2018**, *11, 91–107*, https://doi.org/10.29119/1641-3466.2018.117.6.
[18] Environmental Protection Agency. *Life cycle assessment: Inventory guidelines and principles. EPA/600/R-92/245*. Office of Research and Development, Cincinnati, OH, USA, 1993.
[19] Müller, L. J.; Kätelhön, A.; Bachmann, M.; Zimmermann, A.; Sternberg, A.; Bardow, A. A guideline for life cycle assessment of carbon capture and utilization, *Frontiers in Energy Research* **2020**, *8*, https://doi.org/10.3389/fenrg.2020.00015.
[20] Guinée, J. B. *Handbook on life cycle assessment. Operational guide to the ISO standards*. Springer Netherlands Eco-Efficiency in Industry and Science Ser, v.7, 2002.
[21] De Falco, M.; Iaquaniello, G.; Centi, G. *CO_2: A valuable source of carbon*. Springer, 2013, https://doi.org/10.1007/978-1-4471-5119-7.
[22] Sathre, R.; Masanet, E.; Cain, J.; Chester, M. The role of Life Cycle Assessment in identifying and reducing environmental impacts of CCS, LBNL Report.
[23] Jeon, J.; Bae, C.; Johansson, B. *Effect of hydrogen and DME injection in homogeneous charge compression ignition engine with DME second injection strategy, Institution of Mechanical Engineers – Internal combustion engines: Improving performance, fuel economy and emissions*. Woodhead Publishing, 2011, 157–166, https://doi.org/10.1533/9780857095060.5.157.
[24] International Energy Agency Home Page. Assessed January 24, 2022. https://www.iea.org/news/a-multi-source-multi-purpose-fuel-for-the-future.

[25] Gas Encyclopedia Air Liquide Home Page. Assessed January 24, 2022. https://encyclopedia.airliquide.com/dimethylether.
[26] Pérez-Uriarte, P.; Ateka, A.; Aguayo, A. T.; Gayubo, A. G.; Bilbao, J. Kinetic model for the reaction of DME to olefins over a HZSM-5 zeolite catalyst, *Chemical Engineering Journal* **2016**, *302*, https://doi.org/10.1016/j.cej.2016.05.096.
[27] Pérez-Uriarte, P.; Ateka, A.; Gayubo, A. G.; Cordero-Lanzac, T.; Aguayo, A. T.; Bilbao, J. Deactivation kinetics for the conversion of dimethyl ether to olefins over a HZSM-5 zeolite catalyst, *Chemical Engineering Journal* **2017**, *311*, https://doi.org/10.1016/j.cej.2016.11.104.
[28] Yang, G.; San, X.; Jiang, N.; Tanaka, Y.; Li, X.; Jin, Q.; Tao, K.; Meng, F.; Tsubaki, N. A new method of ethanol synthesis from dimethyl ether and syngas in a sequential dual bed reactor with the modified zeolite and Cu/ZnO catalysts, *Catalysis Today* **2011**, *164(1)*, https://doi.org/10.1016/j.cattod.2010.10.027.
[29] Zhou, H.; Zhu, W.; Shi, L.; Liu, H.; Liu, S.; Ni, Y.; Liu, Y.; He, Y.; Xu, S.; Li, L.; Liu, Z. In situ DRIFT study of dimethyl ether carbonylation to methyl acetate on H-mordenite, *Journal of Molecular Catalysis A: Chemical* **2016**, *417*, https://doi.org/10.1016/j.molcata.2016.02.032.
[30] Kodera, Y.; Kaiho M. Model calculation of heat balance of wood pyrolysis, *Journal of The Japan Institute of Energy* **2016**, *95, 881–889.*
[31] Koponen, J.; Kosonen, A.; Huoman, K.; Ahola, J.; Ahonen, T.; Ruuskanen, V. *Specific energy consumption of PEM water electrolysers in atmospheric and pressurised conditions.* In 18th European Conference on Power Electronics and Applications (EPE'16 ECCE Europe), 2016, 1–10.
[32] Shiva Kumar, S.; Himabindu, V. Hydrogen production by PEM water electrolysis – A review, *Materials Science for Energy Technologies* **2019**, *2(3)*, https://doi.org/10.1016/j.mset.2019.03.002.
[33] De Falco, M.; Capocelli, M.; Centi, G. Dimethyl ether production from CO_2 rich feedstocks in a one-step process: Thermodynamic evaluation and reactor simulation, *Chemical Engineering Journal* **2016**, *294*, https://doi.org/10.1016/j.cej.2016.03.009.
[34] Khan, M.; Saud, K.; Ashraf, K.; Irfan, M.; Ibrahim, S. Curing of concrete by carbon dioxide, *IRJET*, **2018**, *4410–4414.*
[35] Jones, J.; Barton, C.; Clayton, M.; Yablonsky, A.; Legere, D. SkyMine carbon mineralization pilot project, Final Phase 1 Topical Report, September 2011. Accessed June 1, 2021. https://digital.library.unt.edu/ark:/67531/metadc845150/m2/1/high_res_d/1027801.pdf.
[36] Zampori, L.; Saouter, E.; Schau, E.; Cristobal, J.; Castellani, V.; Sala, S. *Guide for interpreting life cycle assessment result*, JRC Technical Reports, European Commission, 2016, ISBN 979-92-79-64114-5.

SI to US units conversion factors

SI units	US units	Conversion factor SI → US
	Mass	
μg (10^{-9} kg)	gr	0.000015
mg (10^{-6} kg)	gr	0.015432
g (10^{-3} kg)	oz (troy)	0.032151
	oz (av)	0.035274
kg	U.S. cwt	0.022046
	lbm	2.204622
Mg (10^{3} kg)	U.S. ton	1.102311
Mt (10^{9} kg)	U.S. ton	1.102311×10^{-6}
Gt (10^{12} kg)	U.S. ton	1.102311×10^{-9}
	Substance amount	
mmol (10^{-6} kmol)	lbm·mol	2.204622×10^{-6}
mol (10^{-3} kmol)	lbm·mol	2.204622×10^{-3}
kmol	lbm·mol	2.204622
	std m^3 (0°C, 1 atm)	22.413584
	std ft^3 (60°F, 1 atm)	836.610056
	Pressure	
kPa	lbf/ft^2	20.885434
	bar	0.01
	mmHg (0°C)	0.145038
MPa	U.S. $tonf/in^2$	0.072519
	kgf/mm^2	0.101972
	U.S. $tonf/ft^2$	10.442718
	lbf/in^2 (psi)	145.037744
	mmHg (0°C)	145.037744
	Temperature	
°C	°F	(°C·9/5) + 32
K	°R	K·9/5
	Length	
nm (=10^{-9} m)	mil	0.000039
μm (=10^{-6} m)	mil	0.039370
mm (=10^{-3} m)	in	0.039370
m	yd	1.093613
	ft	3.280040

(*Continued*)

SI units	US units	Conversion factor SI → US
	Area	
m^2	ha	0.0001
	yd^2	1.195990
	ft^2	10.763910
	Volume	
dm^3	U.S. gal	0.219969
	U.S. qt	0.879877
	U.S. pt	3.141593
m^3	yd^3	1.307951
	bbl (U.S.)	6.289811
	ft^3	35.314662
	U.S. gal	264.172037
	Energy	
kJ	cal	239.005736
	kcal	0.239005
	Btu	0.947817
MJ	kWh	0.277778
kWh	MJ	3.6
	Btu	3412.141286
	Power	
kW	Btu/s	0.947817
	hp (electric)	1.340483
MW	Million Btu/h	3.412141

Index

Note: **Bold** page numbers refer to tables and *italic* page numbers refer to figures.

For Product Safety Concerns and Information please contact our EU representative GPSR@taylorandfrancis.com
Taylor & Francis Verlag GmbH, Kaufingerstraße 24, 80331 München, Germany

www.ingramcontent.com/pod-product-compliance
Ingram Content Group UK Ltd.
Pitfield, Milton Keynes, MK11 3LW, UK
UKHW020206200926
475702UK00004B/638
* 9 7 8 1 0 3 2 3 7 3 5 4 6 *